Energy Systems

Other Books in McGraw-Hill's Complete Construction Series

Dodge Cost Guides ++Series

All from McGraw-Hill and Marshall & Swift

Energy Systems

Robert Carrow

McGraw-Hill

New York San Francisco Washington, D.C. Auckland Bogotá
Caracas Lisbon London Madrid Mexico City Milan
Montreal New Delhi San Juan Singapore
Sydney Tokyo Toronto

Library of Congress Cataloging-in-Publication Data

Carrow, Robert.
    Energy systems / Robert Carrow.
        p.   cm.
    Includes index.
    ISBN 0-07-014019-7
    1. Power resources—Handbooks, manuals, etc.   2. Energy
conservation—Handbooks, manuals, etc.   I. Title.
TJ163.235.C37   1999
696—dc21   98-51760
                                                                            CIP

McGraw-Hill

A Division of The McGraw·Hill Companies

1 2 3 4 5 6 7 8 9 0   DOC/DOC   9 0 4 3 2 1 0 9

ISBN 0-07-014019-7

*The sponsoring editor for this book was Zoe G. Foundotos, the editing supervisor
was Paul R. Sobel, and the production supervisor was Pamela A. Pelton. It was
set in ITC Century Light by Jana Fisher through the services of Barry E. Brown
(Broker—Editing, Design and Production).*

Printed and bound by R. R. Donnelley & Sons Company.

McGraw-Hill books are available at special quantity discounts to use as premiums and
sales promotions, or for use in corporate training programs. For more information,
please write to the Director of Special Sales, McGraw-Hill, 11 West 19th Street, New
York, NY 10011. Or contact your local bookstore.

This book is printed on recycled, acid-free paper containing a minimum of 50%
recycled, de-inked fiber.

To my in-lawed parents, Arthur and Lucille Cannon, who provided many years of help, knowledge, and support. I thank you.

Also, to my wife Colette, my son Ian, and my daughter Justine . . . who all helped to make this project a success. I thank you, also.

Contents

8 Residential Mechanical Systems 137

9 Residential Electrical Systems 173

10 Residential Automation 245

Foreword

What an awesome present and future we have, working together to teach new methods and quality ideas, from using recycled products to saving energy . . . which in the long run, saves you money! As your and my family are so important, we want the best for them. This is important in selecting construction methods. My saying in the construction field is, " . . . we are building your dreams of tomorrow, today!"

Let's reflect on the past, back to the 1960s and 1970s. Sure we still build with some of those methods of that time, but those years have come and gone. Today, we have numerous methods such as geothermal heat pumps and gas-fired, in-floor hot water circulating heat to all levels of the home if desired. That's right, no more cold ceramic tile floors. Here in the 1990s we also have high-efficiency heat pumps and air conditioners, building wrap for less heating and cooling losses, and direct vent fireplaces which are 78 percent efficient. We use insulation with R values of R-38 and high efficiency glass in our windows to keep out the sun's rays and thus reduce our air conditioning costs. The list goes on, but we must always think of the future. We promise to learn and to use new methods and products as they become available.

Having been in the building business since 1976, I have worked with ideas of energy savings for many years and still do. My new ideas of energy savings come from experience, old and new knowledge, trials and performance that these "systems" work. The last ten years, I've been building custom homes in the Pittsburgh, Pennsylvania area, which has a climate that includes all four seasons . . . a real challenge for energy efficient construction. Having been recognized for Best of Show and Best Innovative Design, indicates that people appreciate the effort and commitment to provide them with their dream house.

In talking about people, these are my customers and they are a reflection of me, the builder. Every home built by my company is unique, which is very important for the home buyer. The relationship between me the builder and the home owner is as unique as is the house. Hours upon hours to make their dream home a dream of today. I feel that customers strive to do well at their line of work and it is my job to perform just as well. My customers need to be assured that they are going to receive the "best of the best" when building with me.

I would like to compliment the electrical power companies, locally and around the country for all their support. With their knowledge and help, they have taught me energy savings and how to be a good sense builder through a series of seminars. This helps all of us.

I am privileged and honored to be a friend of the author, Robert Carrow. He is a true friend in that he cares about all that he does and as you read this book you will find that he cares about you, the reader, too. This book, *Energy Systems,* will provide you with first hand knowledge and exciting ideas and skills for energy efficient building. This book will help design that new home improvement or new home by providing the latest ideas of energy efficiency.

Wayne J. Henchar
Wayne J. Henchar Custom Homes
Harmony, Pennsylvania

Introduction

The universe is a collection of energy systems. From the macro-energy systems of all the stars and their planets down to the micro-systems of a single British thermal unit (Btu) leaving your house. If you have gotten to this point of the book then you are serious about energy efficiency. Energy savings has no beginning nor end. We just have to constantly battle the elements, technology, and ourselves when energy use in the home is the issue. The elements because the hot and cold weather along with global climatic changes are more or less out of our control; technology because it keeps changing and forcing itself onto each one of us; and we have to battle ourselves because sometimes man does not have his priorities in the right place. Energy, of any type, is valuable, why waste it? Why not get every bit of savings and efficiency possible out of every material, process, or system?

Energy Systems provides an overview of all those disciplines which make up building construction. It also looks at all those disciplines, each as a system with their associated subsystems and micro-systems, from an energy vantage point. Many tables are provided throughout to use as guides in helping you to make an educated decision. Unfortunately, many issues relative to energy efficiency are not black and white. Therefore it is necessary to perform some basic calculations, make some comparisons, and continue to learn as much as possible about all the energy systems that exist within a new or existing home. With this information, the benefactors are the home owners and home builders of the world.

This book is practical and written for the average home owner so that they can use it. The goal is to provide any information possible to allow for energy savings of any kind within a new or remodeled residence. This book can have a pay back period of one day if it provides just one good energy saving idea. It contains a few formulas but is not "math heavy."

Chapter 1 covers the basics. It defines a system, analyzes heat and its properties, and how electricity is generated. This chapter reviews the fuel and energy sources available to us and provides explanations for power, horsepower, work, and torque.

Chapter 2 on the other hand departs from the technical and addresses the relationships between the active parties in the home building transaction. It is entitled, *Builder-Contractor-Client Relationships*, and details the responsibilities of the home owner, home builder, and the various contractors involved in the process. Various tips and recommendations for a smooth, successful construction project are offered. There are many stages and facets to a building project. The roles and duties of the parties and their interaction is also covered. A milestone/scheduling chart is provided to help keep the project on schedule.

The third chapter, *Energy Savings through Structural Means*, looks at the initial phase of the energy efficient building . . . the structure itself. The size and shape of the structure itself, along with its orientation and framework, is the grandest of all the energy systems in the house. It may make or break the entire energy efficiency component of the project. If various structural and initial design and building characteristics are not checked first then further energy savings may be difficult to incorporate. Isothermal planes, thermal bridges, and dead air spaces are reviewed. A comparison is also drawn between steel construction versus wood. Designing passive solar systems into the home plus a look at geodesic and round house construction are also provided.

Chapter 4, *Modern Residential Building Construction*, gives an overview of conventional building construction and framing techniques. This chapter provides a detailed look at the door and window systems used in residential construction. The topic of insulation is also covered in this chapter. Lastly, several energy saving tips relative to saving electricity, maintenance of equipment, and recommendations are given at the end of the chapter.

Chapter 5 is entitled, *Energy Savings, Codes, Standards, and Safety*. This chapter discusses EPACT, the Energy Star and Energyguide programs, NAECA, and other programs aimed at helping consumers save energy. Demand side management and safety concerns are also covered. Lastly, a fairly complete list of the organizations which set the standards and guidelines for the residential construction industry is given. This list contains the names, addresses, telephone numbers, and a short description of each organization listed. A great deal of free literature and information on energy efficiency and energy savings is available through these organizations.

Various methods are needed to measure energy savings. Chapter 6, *Measuring Energy Savings*, provides examples of calculating heat loss, heat gain, sizing heating systems, and how to monitor energy usage. The home energy rating system is also discussed. Discussion on the phenomena of leaking electricity within the home is also provided. We need to know how much and where energy is being lost in order to control and save it.

Finishing Systems and Energy-Saving Products, the seventh chapter, looks at energy absorption and emissivity of materials and color of components to maximize heat gain. Building wrap, tapes, and sealants are covered. Air lock types and green products are also discussed. The house can look spectacular, all the while conserving energy with its built-in systems.

Chapter 8, *Residential Mechanical Systems*, discusses various types of mechanical systems used for energy efficiency in the home. Efficiencies of heating, cooling and air conditioning systems are covered. Many different mechanical energy systems are reviewed. They include: gas furnaces, hot water heat, refrigeration systems, heat pump and geothermal packages, water heaters, the ductwork system, air changes, house ventilation, whole house fans, centrifugal pumps, piping, and plumbing. There are many different mechanical systems within any one residence and these energy systems use energy, therefore any appreciable energy savings or increased efficiency can come from these product areas.

Chapter 9, *Residential Electrical Systems*, is the longest chapter in this book. It is the longest because much of our everyday life is affected by electrical systems and our households are filled with electrical appliances, gadgets, and equipment. The chapter starts with a review of the average, monthly electric bill and concludes with an in depth look at variable frequency drives which are finding their way on to various pieces of electrical equipment within the home. Basic electricity, some basic calculations (so the average home owner can know what he or she is dealing with, electrically), the hot topic of electric power deregulation is discussed. Topics such as lighting, transformers, and other electrical hardware items are also covered.

Chapter 10, *Residential Automation*, covers the exciting, fast emerging field of home automation. With so many pieces of equipment being furnished with microprocessors, the home begins to look like an electronic, digital jigsaw puzzle. The home automation control system seeks to bring all those electronic components together as one elaborate, working energy system. These systems can provide energy savings control while we're not home or involved. The technology is reviewed and the various components to make the basic system are covered. Electronic wiring types and schemes along with wireless systems are discussed. Lastly, the byproducts of this new technology such as electrical noise and troubleshooting a system are also covered.

No energy systems book could be complete without a chapter on solar energy. Chapter 11, *Solar Energy*, does not disappoint. Solar fundamentals are provided along with information on solar insolation (not insulation). Heat gain and passive solar energy systems for heating and electricity generation are covered.

The final chapter gives a future look at energy systems, trends, and controls for residential use.

Energy efficiency and energy savings are important to every one of us. It is a very logical thing to strive for in our residences. This book should provide an idea or two to help you achieve that goal.

Robert Carrow

Energy Systems

Energy-Saving Systems

Consider this: The United States has a population of roughly 250 million people. The entire earth's population is around 5 billion people. This means that Americans represent 5 percent of the total population on this planet, yet we use the most energy! Our energy consumption is incredible. Look at each household, count the electrical appliances, note the heating and cooling equipment, and throw in an average of two automobiles per family, and this is not as amazing as it appears. The United States is extremely dependent on all types of energy sources. Most humans on this planet do not have the luxuries of air conditioning, lavish audio and visual systems, and automobiles. We had better stop talking about these facts and take measures to correct for our future.

Speaking of the future, this is where all residential design should be focused today. Frank Lloyd Wright maintained that buildings should be designed to be in harmony with their surroundings. Make the structure fit the terrain rather than make the terrain fit the structure. While this concept is important both from an energy efficiency standpoint and from an architectural standpoint, we still must not lose sight of the future, especially when it comes to energy. The fossil fuel resources available today either are not going to be around in the short-term future or else will escalate so high in price that only the rich will be able to afford them. Both these possibilities mean that we are going to have to make some changes in the design and building of our homes today.

In addition, the earth's climate is constantly changing. Even as we battle to try to keep combustion and emissions to a minimum to save the atmosphere, our planet is an object that is constantly changing. We will see several severe storms in our lifetime. We may even be able to perceive a climatic change over the course of our life. The weather is going to be different year to year, and we have to anticipate some of this change as we design our homes for the future. The direct combustion of a material to generate heat is never the most efficient use of a fuel.

Global warming is one of the future possibilities that may have to be considered when designing a home. Global warming, whether by human-generated conditions or just a natural change in the earth's evolution, means basically that the average temperatures over time will rise. Scientists predict that temperatures will rise by one or two degrees over the next half century. This means that any design tables used for heat loss and heat gain calculations should be adjusted accordingly. Even though this rise in average temperature does not seem all that dramatic, a minor change can mean longer warm seasons and shorter cold seasons. Additionally, these temperature rises also may give way to more frequent and more severe storms during the course of a year. Even though conventional modern building construction incorporates many energy-efficient designs and the packages are typically well built to withstand most storms, the home builder and home owner need to consider the changes of the future, especially when trimming or lessening the design due to cost or size issues.

The sun strikes the earth with ultraviolet radiation while it warms the planet. This ultraviolet radiation could cause severe damage to all living organisms on the earth's surface if it were not for the ozonosphere. The ozonosphere is the region in the upper atmosphere as far as 30 miles up in altitude where there are appreciable concentrations of the molecule ozone. Ozone, O_3, is a three-atom allotrope of oxygen in which the molecule has three atoms instead of two. This accounts for the distinctive odor after a thunderstorm or around electrical equipment. However, its major contribution to our planet is its role in absorbing solar ultraviolet radiation. The ozone layer is around 25 miles thick, but the total amount of actual ozone gas is quite small compared with the other gases in the layer. However, there is present-day still enough to absorb the damaging ultraviolet rays.

Many air pollutants, in particular nitrogen oxides, halons, and chlorofluorocarbons (of which many are now banned), can diffuse into the ozonosphere and destroy the ozone. There is even reportedly a hole in the ozonosphere, or periodically a very thin layer of ozone, over Antarctica. With energy-using products comes the need to make combustion to generate the electricity. Likewise, other fossil fuels that are burnt continue to supply more pollutants to the atmosphere. This is why energy efficiency is so important.

For those individuals and home owners who rely on logic to make their decisions, energy efficiency is very logical. If we attempt to find energy savings in every possible system within our residence, then we get rewarded by spending less of our hard-earned income to pay for the what-used-to-be wasted heating oil, natural gas, or electricity. We build in protection from future energy source shortages and price increases. We also make a serious contribution to saving the ozonosphere and the environment. These factors all make taking this serious effort to strive for the best energy efficiency in our residential systems very, very logical (and practical).

What Is a System?

A *system* is defined as an assembly of components with specific structure and intended function. A *residence* is an assembly of components with specific structure and intended function. The two definitions are mirror images of each other. A house is a system; actually, it is several subsystems all working together to form the comfortable cozy building that we call home. Every component within a home has to work as part of the master system. As components interact and come into contact with other components within the residence, the efficiency of that particular subsystem is affected. The goal of the home builder and the home owner is to analyze the interaction of all the components and ensure that energy efficiency is maximized.

A residence has mechanical, electrical, computer, structural, and thermal systems in place that make it much more than just a shelter. When discussing energy systems for residential construction, it is necessary to break each of these subsystems down into their individual elements in order to achieve the very best efficiency possible. Any subcomponent of any system has an efficiency attributed to it. If we strive to maximize the efficiencies of all the pieces of any system, then the system itself will gain an overall better efficiency. This is in essence what manufacturers do to make their assemblies and systems more efficient: They try to make each component as efficient and cost-effective as possible. Especially in an industry where efficiency sells the product, these manufacturers are constantly looking for an edge. This is research and development at its finest. One thing is for sure, we *can* make a product better or more efficient if we set our minds and efforts to doing just that. However, often other priorities get in the way, and many ideas get shelved.

Heat

Cold is the absence of heat, and heat is an energy in transition. We create it, try to retain it for a while, and then make some more. For years we have been trying to find new ways to create heat. The energy question is perhaps the oldest, next to the wheel's invention and what to do with it next. What is the best source of heat, and what are the alternatives? Creating heat is necessary, but there is a finite amount of resources available to create heat for the billions of people on this planet. The amount of heat also has to be finite because most fuel sources will exhaust themselves someday. The debate centers around which fossil fuels will diminish first and when. In order to make better use of the heat that we create, we need to understand why it is an energy in transition. We also need to know how and why heat migrates so that we can contain and control it. This is the essence of energy efficiency.

Heat transfer happens by three primary methods. One method is called *conduction,* the transfer of heat energy through a material by the motion of internal, adjacent atoms and molecules. Heat transfer by conduction provides no observable motion of the material. In metallic solids the motion is of unbound electrons, whereas in liquids it is the transport of momentum between molecules. In gases, it is the diffusion of molecules containing heat; this also creates the random motion of the gas molecules. Heat transfer by conduction has even undergone elaborate mathematical study by J. B. Fourier, a famous French mathematician.

The Fourier equation basically calculates the thermal conductivity, k, through a thickness of material while factoring in the temperature difference between the two surfaces and the square footage. This equation is the basis for many of the heat loss and heat gain calculations done for energy efficiency. The equation is as follows:

$$Q = \frac{k}{\Delta X} (A) (T_2 - T_1)$$

where Q is the heat flow, k is the thermal conductivity, ΔX is the thickness of the material, A is the area of the material, and T is the temperature. This can be seen graphically in Fig. 1.1. High thermal conductivity means that a material is a good thermal conductor, and low thermal conductivity means that a material is a good thermal insulator. Heat transfer by conductivity also can be likened to a system. There are many components to the heat-transfer equation, and all facets of this "thermal system" have to be analyzed.

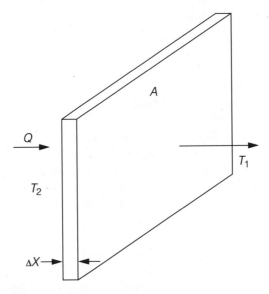

■ **1.1** *The Fourier equation at work.*

Another method of heat transfer is by *convection*. Convection and convection currents arise from the mixing of fluid elements. Whenever there are differences in density, then there is the chance for natural convection. Convection is also a mechanism by which we can deliver and distribute heat to a residence. It is a mechanism because mechanical means are used to move heated or cooled air around a residence. The same can be said for moving fluids around a residence. The negative contribution of convection to a residence is the phenomenon of allowing heated or cooled air to be convected out of the house through cracks and tiny openings. All this portrays residential heat transfer as another energy "microsystem."

The last type of heat transfer is *radiation*. Heat transfer by radiation is done by electromagnetic waves. All materials are said to radiate thermal energy, and whenever radiation falls on a second body, it will be either transmitted, reflected, or absorbed. Absorbed energy appears as heat in the body. The ultimate example of this is the heat energy radiated to the earth each day by the sun. The earth absorbs, reflects, and transmits this thermal radiation, as it should. This entire life-sustaining process is the most elaborate energy system we could classify.

Unwanted Heat

There are three major sources of unwanted heat in your house during the summer: heat that conducts through your walls and ceilings from the outside, waste heat that is given off inside your house by lights and appliances, and sunlight that shines through the windows. All three methods of heat transfer are at work here: radiation, convection, and conduction. To reduce heat gains through walls and ceilings, you can add insulation and seal up cracks to reduce air infiltration. Ventilating your attic can be an important measure to reduce significant heat buildup during the summer that finds its way into your home. The remainder of this book will look at all aspects of energy efficiency and energy savings for residential construction.

Renewable-Energy Sources

Renewable energy sources are those which do not require the combustion or destruction of the fuel. Fossil fuels are not renewable energy sources: When they are gone, they are gone; they will not be able to renew at the rate at which we are using them. When one considers that it takes millions of years to form oil, gas, or coal from dead and decayed plant and animal material, it is obvious that the inevitable must happen: There will be no more oil, coal, or gas. This may happen 3000 years from now, or it might happen in 50 years. It all depends on our rate of use and our energy waste. Therefore, we can and will use alternative sources for energy, and they will continue to be developed. The following paragraphs describe the renewable energy sources in use today.

One type is *geothermal energy,* which is generated by converting hot water or steam from deep beneath the earth's surface into electricity. Any time thermal energy is gathered from the earth's groundwater or geology, the term *geothermal* is given it, and this is the basic premise for geothermal heat pumps. Geothermal plants that convert hot water or steam from within the earth are not that common yet. However, geothermal plants emit very little air pollution and have minimal impacts on the environment.

Another renewable energy source is called *biomass.* Organic matter, called *biomass,* can be burned in an incinerator to produce energy. In newer facilities, the biomass is converted into a combustible gas, allowing for greater efficiency and cleaner performance. Biomass resources include agricultural, forestry, and food-processing by-products, as well as gas emitted from landfills.

Water dams provide what is called *hydroelectricity* by guiding the water down a chute and over a turbine at high speed. This constant barrage of water over the turbine produces electricity, which is then supplied to the electrical power grid. Smaller-scale hydropower, which is considered to be less than 30 MW, is classified as a renewable energy source, whereas large-scale hydropower is not. Although hydropower does not produce any air emissions, large dams also must address environmental issues such as flood control, water quality, and fish and wildlife habitat. Therefore, only small hydropower is considered "green," i.e., renewable.

The *wind* is a source of clean, renewable energy. Wind turbines, sometimes called *windmills,* use strong, steady winds to create electricity. Fields of windmills in areas of the country where the winds are strong and steady use an induction motor to produce the electricity. As can be seen in Fig. 1.2, the wind forces the blade of the fan to rotate, which causes the electrical motor to rotate. Induction motors, when given electrical power, consume energy and move their loads. This is called *motoring.* Whenever an induction motor is driven by its load, it is called *regeneration.* The motor now becomes a generator—a generator of electricity. This electrical power is harnessed and supplied to the power grid for consumption. Wind power emits no pollution and has very little impact on the land. Wind energy can be produced anywhere the wind blows with consistent force.

Solar energy systems are another type of renewable energy source. They are perhaps the most prominent today because many residences and businesses have incorporated some form of energy efficiency into their overall building design. Chapter 11 also discusses solar energy systems as they pertain to residential construction. Basically, the sun's radiation is used directly to produce electricity in two ways, photovoltaic systems and solar

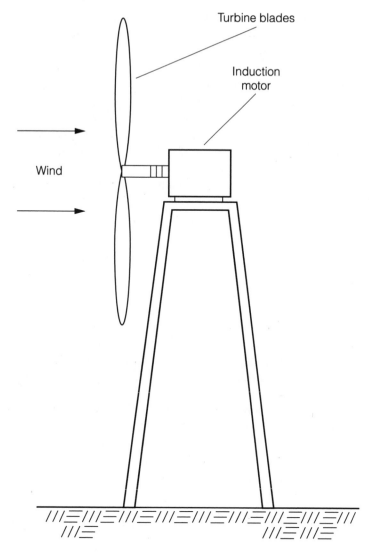

Turbine blades

Induction
motor

Wind

■ **1.2** *A wind turbine with induction motor/generator.*

thermal systems. *Photovoltaic systems* change sunlight directly into electricity. *Solar thermal systems* use the sun's energy to heat a fluid that produces steam, which then turns a turbine and generator. Smaller-scale versions of these can be used residentially, but many homes use the radiation from the sun in some way or another.

Nonrenewable-Energy Sources

Nonrenewable energy sources are classified as such because of their impact on our environment. They consist mainly of the fossil fuels. As discussed earlier, fossil fuels are approaching a time when they cannot be replenished to meet the demand. Additionally, some hydropower dams that provide electricity by guiding water down a chute and over a turbine at high speed have dramatic effects on the environment. Small-scale hydropower, less than 30 MW, is considered renewable, whereas large-scale hydropower is not. Although hydropower does not produce any air emissions, large dams also must consider environmental issues such as flood control, water quality, and fish and wildlife habitat. Therefore, while large-scale hydropower is clean, it is not kind to the environment with regard to the ambient surroundings. Hydroelectric power will continue to be developed as an alternate energy source for the future, however. Large-scale hydroelectric power is somewhat regional. Large bodies of fast-moving water are the prime locations, and currently, about one-quarter of the power generated in California is from hydroelectric power plants.

The bulk of the nonrenewable energy sources is covered by coal, natural gas, and oil. Coal is developed from decaying plants and becomes a black or brown rock. Strip mining or underground mines are built to extract coal. These mines cause severe erosion, leaching of toxic chemicals into nearby water sources, and loss of habitat. About 65 percent of the sulfur dioxide, 33 percent of the carbon dioxide, and 25 percent of the nitrogen oxide emissions in the United States are produced by coal-burning plants. These emissions contribute to global warming, acid rain, and various threats to health. Oil burns cleaner than coal but still produces large quantities of pollution per unit of energy produced. With most of the oil in the United States already extracted (this proves that we have started to run out of fossil fuels in certain areas of the world), future dependence on oil would require an increase in imports. Natural gas is inexpensive and environmentally benign. These two factors have greatly increased the use of natural gas to generate electricity. Natural gas does produce air pollution, but not nearly as much as other fossil fuels. Although natural gas reserves will last for many decades to come, they are finite, and with scarcity, the price of natural gas will rise.

The last of the nonrenewable energy sources is nuclear power. Nuclear power comes from splitting uranium or plutonium atoms. Nuclear power poses grave risks to both human health and the environment, even though there are no emissions. A nuclear power plant accident could spew radioactive materials into the atmosphere, causing catastrophic damage. Remember the Chernobyl disaster. While a nuclear accident is remote, safely storing nuclear waste is a real problem. Nuclear fission creates materials that will remain dangerously radioactive for thousands of years. It is impossible to ensure that any underground storage site will be safe for such a long period of time.

Changing Electrical Power into Mechanical Power

Inside the home there are conversions going on, and these conversions of electrical energy into mechanical power are making things move. Electric current changes to watts and ends up as torque moving air or fluids. Much of the mechanical power transmission equipment now has some electrical control. Home owners must have a good understanding of what is taking place, mechanically, in order to assimilate their electrical needs. The motorized equipment in the average home is a large consumer of energy. Understanding the basics of energy, power, work, and torque are important in understanding how we might save energy in our homes. Additionally, with so many appliances and other household equipment being electrically and electronically controlled, it is vital that we begin to know as much as we can about the power mechanics, or power transmission, systems within our kitchens, laundries, and basements. We cannot always call the specialized technician for every "blip" or "burp" that our equipment or appliances make when they're not working properly.

More than likely, anything moving in the house is being driven by an electric motor. This movement by these prime movers provides some amount of work. *Work,* in its basic definition, is a force acting through a distance. Work is equal to a force times distance ($F \times D$). A force is a push or pull that causes motion of an object. The object moves in a straight line in the direction of the force applied to it. *Power* is the amount of work done in a period of time and it is usually expressed in horsepower (hp) or in watts (W) for electrical power. Figure 1.3 gives an example of work and how it is expressed. It shows that if 15 lb is moved 10 ft, then 150 foot pounds (lb•ft) of work has been expended.

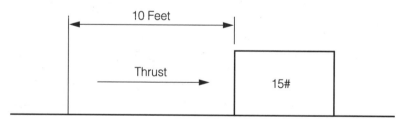

■ **1.3** *Work = 10 ft × 15 lb = 150 ft•lb.*

As we look at work—and further at force, power, and torque—many of these terms seem to be interchangeable. The fact is that many times they *are* substituted in dialogue to get a point across. The bottom line is that if everyone using the terminology had a good, basic understanding of each term, then there might be less confusion and just maybe a few less problems with respect to motors, equipment, and appliances. For instance, as we will see in more detail, torque and work are very similar. Work is a force times a distance, whereas torque is a force times some radius. Close, but yet different. It all depends on where the term is applied and what is trying to be accomplished.

The device used to convert electrical energy into torque is the electric motor. The point in the system where this distinction is made is usually at the coupling off the motor shaft. Torque is transmitted from the motor shaft to the fan blade or pump impeller in order to achieve work. Somehow terminology and usage have created confusion regarding horsepower, and this is a point worth discussing. In the late eighteenth century, James Watt, of Scotland, determined that 1 horsepower was equal to 33,000 foot pounds of work in 1 minute. This is equivalent to the amount of power required to lift 33,000 pounds 1 foot in 1 minute (Fig. 1.4). A horse, being the working beast in those days, was selected as the "prime mover," and these values were attributed to the amount of work a horse could do. Horsepower can be derived from the following equation, where force is F, D is distance, and t is time:

$$\text{Horsepower} = \frac{F \times D}{33,000 \times t}$$

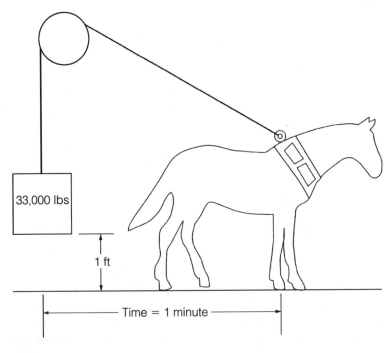

■ **1.4** *Horsepower.*

The term *horsepower* has probably caused more costly undersizing when it comes to variable-frequency drives and motors than we can fathom. The fact of the matter is that when sizing the foot-pound requirements of a given application, we are really concerned about torque. After that, it is important to then look at the electric current requirements of the application and how they relate to torque. The electrical equivalent of 1 horsepower is 746 watts. To find a value for horsepower (hp) when other values are known, the following base formula may be used:

$$\text{Horsepower (hp)} = \frac{\text{torque (ft·lbs)} \times \text{speed (rpm)}}{5250}$$

Torque is defined as a rotating force. Further refinement results in the definition that any twisting, turning action requiring force is *torque*. As is shown in Fig. 1.5, torque is the product of some force times a distance, or $T = F \times r$. If a force F is applied to the lever arm at a distance equal to the radius r shown, then a resulting torque is produced. Derivations of this formula can give us the amount of work as a torque acting through an angular displacement. There are mainly two types of torque, static and dynamic. A rotating apparatus will exhibit dynamic torque. It is in a state of constant movement, correction, and change. The universal formula for determining the torque of a rotating piece of equipment is

$$\text{Torque} = \frac{\text{horsepower} \times 5250}{\text{revolutions per minute}}$$

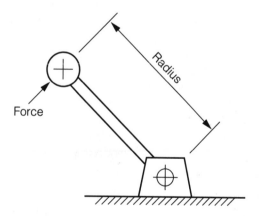

■ **1.5** *Torque.*

There are several derivations of this formula as follows:

$$\text{Horsepower} = \frac{\text{torque} \times \text{speed}}{5250}$$

where torque is in foot pounds, or

$$\text{Horsepower} = \frac{\text{torque} \times \text{speed}}{63,000}$$

where torque is in inch pounds, and

$$\text{Speed (rpm)} = \frac{5250 \times \text{horsepower}}{\text{torque (foot pounds)}}$$

$$\text{Speed (rpm)} = \frac{63,000 \times \text{horsepower}}{\text{torque (inch pounds)}}$$

These are merely the same formula restated to find the proper unknown value when the others are known. Speed, which is sometimes referred to as revolutions per minute (rpm), also can be shown in formula form as N. At 1750 rpm, it is a rule of thumb that 3 ft·lb of torque equal 1 hp. Also, 5250 is a constant, and it is for use with foot pounds of torque. Likewise, the 5250 constant is changed to 63,000 when torque is in inch pounds. A speed, torque, and horsepower nomogram is provided in Fig. 1.6. Displayed is the torque at 1750 rpm and 1 hp, namely, 36 in·lbs or 3 ft·lb of torque. The nomogram can be used to find the third unknown if two of the other values are known. Use this as a guide.

Because your monthly electricity bill has all the previous month's charges for the electric current you used, you must be concerned with electric motor torque. Electric motors operate on electric current, which is nearly proportional to motor torque. Torque is definitely a more appropriate value to use to determine what size prime mover is needed to move the load accordingly. Once the torque value is known, then a determination of which motor to use is much closer. So often we ask a supplier for a certain horsepower motor, many times, say, because this is the size that was on the furnace. Quite possibly, however, a smaller, more energy efficient model could be used. The problem is that motors are sized by horsepower, and this can be potentially dangerous. The better scenario is to provide the equipment supplier with speeds and torque requirements at those speeds so as to size the motor for duty cycle and complete heat dissipation.

A torque wrench is sometimes used to tighten nuts onto a seated surface. This device measures the amount of force in inch pounds. This same device can be used to measure the torque requirements of a particular shaft that moves a fan or pump. This evaluation is good for any rotating component and will give a good indication of the electrical requirements for that application. Also remember that a motor is a stupid device. If a motor cannot turn because its load is too great, then the current to the motor will increase to try to move that load until the supply is shut off. This is why elec-

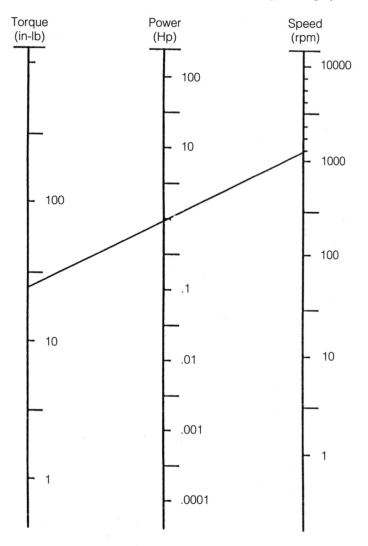

■ **1.6** *Speed-torque-horsepower nomogram.*

tronic overload protection devices or even variable-frequency drives or fuses always should be implemented into a motor system. Alternating current (ac) and direct current (dc) motors should be protected. If a motor does not want to turn, then the motor may be undersized, and the torque output from that motor may be inadequate to perform the application. Eventually, the motor will burn up.

Knowing just a little more about the electrical, mechanical, and motorized systems within our homes can save money in more areas than energy. Selecting the proper motor, recognizing an overloaded condition, and knowing how to reset and keep equipment running will pay dividends to the home owner in the long run. It will pay for several reasons. First, the better understanding of how much and where energy is going throughout the house will pave the way for energy management. Second, keeping those high-priced service personnel out of the house for minor mainte-nance and service needs of the appliances definitely will save you money. Last, replacing damaged components due to neglect or abuse can be greatly minimized as the home owner and home builders better understand all the "working" systems within the residence.

Builder-Contractor-Client Relationships

Clients have a vision of what they want, but they do not always know all the energy systems options. The builder must be able to convey these ideas effectively and sell them to the client. This exchange of ideas and concepts is also critical between the subcontractors (mechanical, electrical, etc.) and the general contractor. This chapter provides an in-depth look at these relationships. The roles of each participant are examined, and client benefits are detailed. Key energy-related sales points and strategies for securing projects are given.

The builder-contractor- home owner relationship to each other and to the residence is shown in Fig. 2.1. They all have a vested interest in the design, construction, and completion of the home, yet they all have other day-to-day interests as well. The contractors are usually juggling many projects with possibly many home builders and at other job sites. The contractors have narrow windows for starting and finishing their work, and if there are any delays (inclement weather, personnel and materials shortages, or late deliveries), then the juggling act gets tougher. The home builder, on the other hand, may have several buildings being constructed at the same time. Thus the builder has to juggle many job sites and many subcontractors. However, to the client, the home owner, this project is the only one of interest. The home owner is only concerned with the timely, proper completion of his or her home so that he or she can schedule the moving of furniture, transfer of schools and records, and all the things necessary when moving from one residence to another. It is said that the buying, building, or selling of a house is perhaps the most stressful occurrence in a person's life. The home builder and the contractors have to recognize this stress in the home owner and be willing to work hard to make the process easier.

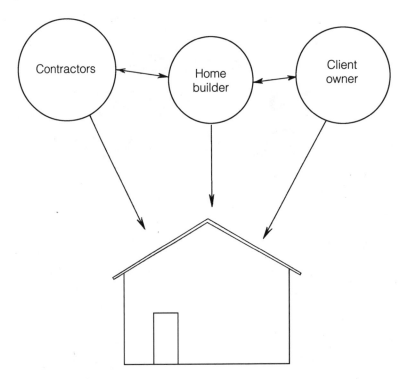

■ **2.1** *Relationship between client, home builder, and contractors.*

The Builder

The home builder has to bring something to the party other than a finished residence. The home builder has a relationship with the clients that can be rather intense. As the clients make their selection as to the person who will build the house of their dreams, those clients are looking for someone they can trust. The clients, usually the husband and wife of a family, are entrusting their future dwelling and shelter along with their hard-earned money to someone they usually do not know. They may have found this particular builder by word of mouth or by calling the chamber of commerce in the region of interest. The selection of a home builder is an interview process, and the home owners are looking for certain qualities in that person to whom they will entrust so much.

The home builder has to be experienced and knowledgeable in his or her chosen profession—building. The home builder must know and understand the federal, state, and local building codes intimately. Other personal qualities that the clients are looking for include honesty, integrity, and the ability to be a good listener. Listening is the first phase of good communication, and if a builder, or any person for that matter, listens and fulfills commitments from those discussions, then the project has a very good chance to succeed. Ultimately, the home builder has to be very energy savings aggressive. Energy efficiency is a prime consideration today and in the future. Keeping up with the latest trends will make the builder a force worth listening to. Many of these qualities are shown in Fig. 2.2.

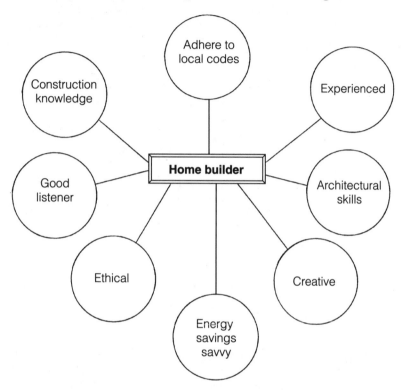

■ **2.2** *The qualities a home builder should possess.*

The home builder, besides having to possess all the qualities described earlier, also must provide most of the documentation. From design and planning meetings with the clients, the builder should prepare floor plans, wall sections, elevations, and even renderings of the house for review. A bill of materials also should be furnished. The home builder is responsible for the overall scheduling of all work and materials along with getting any approvals and testing done with local authorities and inspectors. The home builder also must be able to present these documents accordingly.

From an energy-efficiency standpoint, many home builders fall somewhat short. They are content with a routine building practice that works and may not show the proper interest in energy savings to the client (because they are not paying the electric and utility bills 5 years down the road). A home builder who is knowledgeable about energy savings will continually interject energy efficiency into the design of the home as it evolves. Such a home builder will be able to perform heat loss calculations, pinpoint where energy savings can be obtained, and also use various software packages that address energy savings. The home builder actually can have the biggest impact on the energy efficiency of the house. The home owners usually are not technical enough, and the contractors do not get that involved. Therefore, the home builder can do much more than the state and federal guidelines mandate when it comes to building an energy-efficient home.

Of course, none of the aforementioned services and goods come for free. There is a fee for all professional services, and this is being provided in this transaction—a professional service (plus a new house). While the home builder must properly bid and sell the project to the prospective clients, total price of the project is always a concern for both parties. The home builder is entitled to make a fair profit and be paid for the additional services he or she provides. The clients, on the other hand, want a low price and many times want changes along the way for free. This is why it is commonplace to agree on a contracted amount, have payment schedules based on milestones, and discuss how changes will be handled.

The Contractor

A *contractor* is a person or company that contracts to do or supply something, especially one whose business is contracting work in any of the building trades. This implies many disciplines, and there are. Even the builder is often referred to as the *general contractor.* However, there are a plethora of different individuals and companies that will help to make a construction project a success. This list will only grow in the future as new technologies spawn new disciplines for the residential market.

The contractors that can perform work on site and off site range from electricians to water heater suppliers. As you will see in the following chapters, there are several types of systems incorporated into any one, single residence. Factor in energy savings and energy efficiency, and there are more. There are electrical contractors and mechanical contractors. There are refrigeration, heating, and cooling specialists. There are the plumbers, carpenters, brick and masonry personnel, concrete specialists, control wiring contractors, carpet layers, drywall and roofing companies, door and window people, kitchen suppliers, and the hundreds of other possible providers of an appliance, piece of equipment, or service required for the home. All these individuals or entities are contractors.

The contractors typically are selected, used, and paid by the general contractor or home builder. The home builder, if experienced, usually has used the same contractor for a select piece of work over and over from project to project. Home builders tend to stick with success, and when they find a contractor who performs acceptable work in the time period required, then that relationship continues. Not too often do the home owners get to select the contractors for major house elements (e.g., electrical, mechanical, masonry, HVAC, carpentry, etc.). However, with the secondary disciplines involving appliances, some heating and cooling equipment, types of doors and windows, and so on, the home owners can request a specific contractor. This is another reason why all parties must communicate with each other. In addition, with the growing high-tech component to be integrated into residences being requested by home owners, the home owners can work more closely with these contractors on a direct basis. However, as a rule, the home owners do not get too involved (even minor communication) with the general contractor's subcontractors.

The house is specified and designed, and construction is to commence. The home builder has selected and scheduled many contractors to show up at the site and perform their functions. The home owners are often going to watch their "dream home" be built. All too often they see something they want changed or do not like. They cannot tell the contractor or force the contractor to make the change. The home owners have to go through the home builder first. This controls the content and quantity of changes and how they may or may not affect the overall building's cost. This is all the more reason to define as much detail as possible in the specifications and drawings.

The home builder has to convey the home owners' wishes to the contractors or subcontractors. Again, this is another reason to have things documented, drawn up, and specified. The home builder has to receive bids on work based on something. That something is the building plans and specifications that he or she has prepared for just this purpose. There has to be a binding, legal agreement between the home owners and the contractors. This helps to ensure that before all parties sign, they understand the scope of the work and the extent of their services.

The Client

The client is, in essence, all of us. We all have to live somewhere, and that place is a residence or house. Some call it home. That is, all of us are home owners and have to deal with energy savings, whether our profession is architect, engineer, technician, contractor, or home builder. Energy efficiency is important no matter where you fit in this "food chain." However, from the client's point of view in a transaction such as the construction of a residence, there is a distinct relationship to the home builder and the contractors—the client is the customer.

Whenever money, goods, or services are given in exchange for anything, there is always the potential for dissatisfaction and confrontation.

Everyone has a different interpretation of such terms as *good* or *acceptable, finished* or *complete,* and other terms that can lead to disagreement between two parties. What might seem like a good and finished project to one person may not be acceptable to another, and if this other person has to make payment for the project, then we have the basis of a dispute. This is why the first step must be to open "extremely good" communications between the home owners or clients and the home builder.

Communication occurs in many forms. Verbal communication is fine, even if the trust between the individuals is solid, but it is all right to get things in writing. People accept the fact that there has to be some legal content to all transactions these days, and with this usually being the biggest transaction in a home owner's life, there has to be written, documented, and authorized paper between all parties. Besides, in writing up agreements, specifications, drawings, and contracts, all parties begin an exchange of ideas and actually "design as we go." With energy efficiency as a theme, the home builder, home owners, contractors, and any other involved parties can contribute to a successful project.

One important role of the home owners is that they know what they want (in their minds) and must be able to convey that in detail to the individuals building their home. It may be perceived as burdensome and time-consuming, but the writing of a complete specification for as many elements as possible that will go into the house is recommended. Details regarding things such as materials, labor, appliances, electrical and mechanical equipment, and so on will go a long way toward getting what you want and settling any future disputes. Home owners do not have to be professional specification writers, nor do they have to be very technical. Willingness to do some research and follow some basic guidelines will allow the average home owners/clients/customers to get their points across well. Sketches can be drawn by the home owners to graphically show concepts, and these can be done during one of the several meetings between the home owners and home builder (remember to communicate and do it often) to further get the point across. Leave the professional drawings to the home builder, which he or she can draw from your sketches, and then you can approve their drawings.

As for guidelines, there are many agencies and standards organizations to which home owners can turn. Ironically, these are the same agencies and organizations that home builders use, so the information should be consistent. Many of them are listed in Chap. 5 of this book along with a description of each. Furthermore, this entire book can be your guide to specifying items and systems for your house as they pertain to energy savings. This book covers all the disciplines by which architects and builders categorize their professions. They are listed in Table 2.1. These are the 16 architec-

■ **Table 2.1 Architectural Specification Sections**

Section Number	Category
1	General planning and design data
2	Foundations and sitework
3	Concrete construction
4	Masonry construction
5	Metals
6	Carpentry
7	Thermal and moisture protection
8	Curtain walls, doors, windows, glass
9	Finish materials
10	Specialties
11	Equipment
12	Furnishings
13	Assembled construction
14	Elevators and conveying systems
15	Mechanical
16	Electrical

tural sections that professionals go by. If the home owners use this list as a guide or outline, then a fairly complete specification can be presented.

Another tool that home owners should use is a milestone or scheduling system between the builder and themselves. The entire schedule for building the house should be laid out in the design and final planning stages, during one of the several meetings between the clients and the home builder. An example of a milestone chart is shown in Fig. 2.3. By both parties having the same schedule with dates, it will be easy to see if the project is on schedule or not. In this way, if the project begins to fall behind schedule, adjustments can be made. Incorporate milestone meetings with the builder, on site, throughout the project (two, maybe three, depending on the size of the house).

Energy-Savings Responsibility

As the project evolves during the design stage, the home builder and the home owners exchange ideas on energy savings and energy efficiency. The home builder may run off several iterations of energy calculations before a full scheme is decided on. Most often the calculations do track the promised savings, but what happens if they do not? The home builder obviously has the upper hand (in most builder-client relationships) in knowing what can be done to maximize energy efficiency in the construction of the new residence. Most home owners are not aware of building wrap, slab

Milestone and scheduling chart				
Work or occurrence description	Projected date	Actual date	Payment required	Reason for delay (if any)
Final design acceptance				
Initial testing/approvals				
Excavation/sitework				
Footers/slabs/drainage				
Masonry/concrete work				
Subfloor system (meeting)				
Rough framing				
Rough roofing				
Plumbing				
Electrical				
HVAC (Meeting)				
Exterior finish				
Interior finish				
Clean up				
Landscaping				
Final inspection				
Final acceptance				

These milestones agreed to on this date by:

_____   _____
Home builder   Home owner

■ **2.3** *The milestone scheduling agreement.*

insulation to minimize edge losses, how to orient the house, where to place windows, where to place heating registers, how to route and insulate refrigerant piping, or why a gas pack might be better than a heat pump for a particular installation, and the list can go on and on. The home builder lives in the residential construction marketplace and knows much more about energy savings than does the average home owner/client.

Therefore, the home builder can provide the most assistance in determining the energy-savings schemes that get incorporated into the house. However, home owners have to get educated. Energy efficiency is not a do it once and it's over routine. Every year energy savings and energy efficiency take on new meaning. An energy-saving practice that stood for 20 years may no longer be acceptable. The home owners have got to deal with these issues for as long as they live in the home. Granted, incorporating as many sound energy-efficiency procedures into the initial construction as possible definitely will allow the home owners to rest on their laurels for awhile, but energy costs will continue to escalate, and sources of energy may change. It is important for home owners to do the necessary homework and educate themselves concerning energy-efficient methods, products, and systems for residential use. This should be done regardless of who is responsible. The more home owners know, the better off they will be. This knowledge also can be instrumental in productive early design meetings with the home builder.

To Use or Not to Use a Builder?

Unless the potential home owner has the architectural and construction skills necessary to build a house, the answer to this question is yes, use a builder. Any new construction today almost demands the use of a builder and therefore having to select a good one. Attempting to be an "independent general contractor" at the outset may seem attractive, but rarely are substantial dollars saved in doing so. The builder does this kind of work day in and day out, and a good builder can make all the difference in whether or not a house is energy efficient and continues to pay energy dividends throughout its duration on earth. The curve shown in Fig. 2.4 illustrates how a builder will affect a residence from both an energy-savings standpoint and the project's overall final cost. Doing it without the aid of a good builder can be both costly and less rewarding if the do-it-yourselfer is not up to the task, knowledgeable, and well versed in all the energy-efficiency disciplines involved with the residential construction industry.

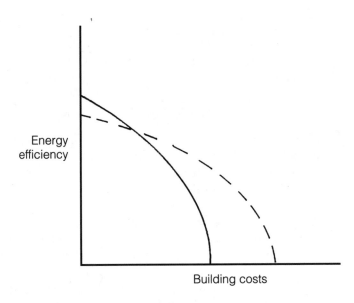

■ 2.4 *The relationship between using a builder and not using a builder with respect to overall house energy efficiency and final project costs.*

Energy Savings Through Structural Means

The energy buck starts here, and it can stop here, too! Sounds a little contradictory, but some foresight when building can "allow" for the energy savings later! If you begin with the structure, all else falls into place. Whether or not you have to make a structural change because of energy savings or are designing new, this facet of construction is the foundation for most of the future energy conservation that one will achieve. The shape of the structure, the framework, the size of the structure, and many minor structure-related construction issues play a role. Understanding the elements of structures, knowing where and why heat losses occur in a structure, and viewing structures and homes as systems will aid in achieving innate and dynamic energy savings for the life of any structure.

In keeping with the theme of this book, another definition of a *system* is an orderly way of doing things. Building a home has to be done this way—as a system. It is *the only method,* and while structural considerations do not necessarily conjure up the thought of a system as we may think of it today, they certainly play a major role in those "electronic systems" later! However, energy savings through structural means is not based solely on the heat-loss savings due to heat transfer from one temperature environment to another. Efficiently routing electrical wire through a house with minimal junctions is important. These sometimes can make for electrical heat losses if the terminations at these points are not solid. Designing the structure so that wiring can be run with few interruptions is important.

Moreover, today's newer bundles of wire may contain several conductors and have to be routed while the house is under construction. Low-voltage control wire is now being "pulled" with normal 115-V ac wire so that electrical outlets can be programmable. These bundles are discussed in Chap. 9 but have to be considered at the outset because they cannot be pulled easily, if at all, later when drywall is in place. These bundles also may predetermine the type of studs used in the walls. Wood framework will have to have larger diameter holes drilled into it to accommodate the larger wire

25

bundles. Metal studs typically have prepunched holes that are large enough to handle these bundles.

Many structural issues must be discussed between the builder and the buyers at the outset. Even though an owner may not be interested in or have much knowledge about structure and framework, it is important to discuss all matters, particularly energy, that can possibly affect the superstructure. This has to be done because, once committed, there may come a point in the construction cycle where *no* changes can be made whatsoever.

Heat is energy in transition. Our job is to not continue to create more of it but rather to impede its movement—hinder the transition. Heat energy is like a trapped mouse; it is trying to find a way to escape and usually will do so. Our job is to contain it. We have to start with the physical structure. The fewer openings, thermal bridges, and isothermal planes, the better. Are these terms and concepts that home buyers should be made aware of? Most certainly.

Energy Considerations

Annual weather conditions, the slope of land, building envelopes, and many other factors interact with any structure. The superstructure, framework, size, shape, and foresight will dictate what we can achieve with poststructural energy systems. As will be seen in subsequent chapters, all facets of residential construction can carry an "energy" component with it. Final structures are several components working in unison to provide a long-lasting residence but also save valuable future dollars on energy costs that surely will increase. Designing in where necessary, yet practical, a nickel spent now will reap inherent dollar savings later. In actuality, it will most likely cost more than a nickel to build in energy savings, but if this is done properly, the costs should not escalate dramatically. After all, the residence requires a sound foundation, solid framework, and tight construction anyhow.

The size and shape of the structure are the two most influential factors. Since heat losses are calculated over the total square footage of the walls, roof, doors, windows, and other surfaces, it is obvious that the larger the structure, the more surface area is available for heat loss. Figure 3.1 shows the basic mathematics concerning this. Even with size as an issue, extra usable living space is always desired; we tend to run out of whatever space we have! A well-thought-out design concerning the residence's usable, livable space versus actual square footage, however, is possible without total sacrifice.

Likewise, the shape of the structure is critical. Cold winter winds (Fig. 3.2) can pound on a conventionally framed square or boxlike residence with

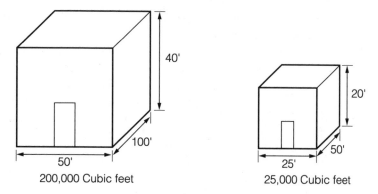

200,000 Cubic feet   25,000 Cubic feet

■ **3.1** *As linear dimensions double, the volume increases eight times.*

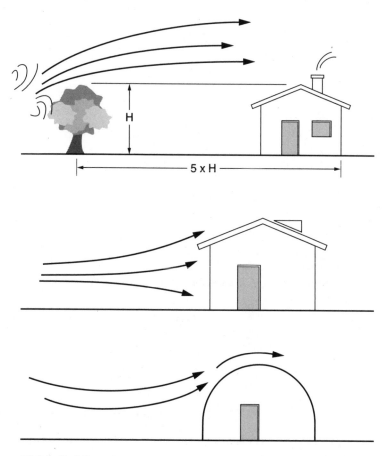

■ **3.2** *Building shape and surroundings relative to wind. The maximum distance of wind-blocking trees from the house is five times the tree height. Square, boxlike shapes do not shed the wind as do rounder shapes.*

such force that extra air changes have to be factored in due to infiltration. Streamlining the structure with rounding effects and giving strong winds proper outlets and reliefs can be built in without much extra cost and effort. Wind loading also has to be weighed in the design of the residence. Northern homes have to incorporate steeper pitches to their roofs due to snow loads, yet the steeper pitch can present a more severe wind load and further contribute to heat losses. A comparison is shown in Fig. 3.3. The style of roof also contributes to the streamlining of the residence or lack thereof.

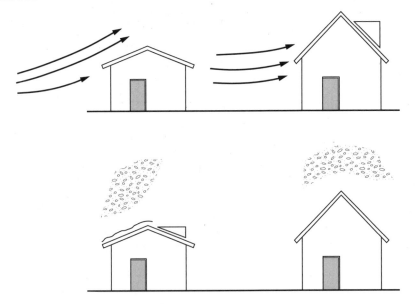

■ **3.3** *Flatter roof pitches allow wind to skim over, but snow loads accumulate. Steeper roof pitches keep snow off the roof, but wind loads are much greater.*

How much of the house should go below grade? The earth surrounding the house acts as an excellent insulator around those components which it covers. The temperature of the ground depends on several things: the part of the country one is building in, the actual composition of the earth (mineral, water, and other elements), the density of the ground, and the time of year. The temperature tends to remain fairly constant, especially at depths 36 inches below grade, ranging from 50 to 60°F (10 to 15°C). This means that the heat transfer across these building surfaces is calculated around a much milder temperature difference from inside to outside.

Just as the residence's structure and design can yield future energy savings, one also must understand what phenomena are causing heat loss in a structural system. Heat losses within a building are usually calculated by the parallel-path approximation method. This method assumes that heat flows straight through a wall or roof. The reality is that heat takes the path of least resistance and moves along a wall until a suitable path of lower resistance (R value) is found.

Isothermal Planes

Energy tends to move along planes parallel to material surfaces. As is shown in Fig. 3.4, heat energy seeks to move along a plane "looking" for a way to transfer itself through that material. Not to give heat any intelligence at all, but it will not pass through a region of insulation. It will continue to move along the isothermal plane until a thermal bridge is found, an opening or crack is found, or the insulation runs out. One property of physics and fluid mechanics is that heat will remain in a fluid (water) or gas (air) until it can be exchanged via thermal conductivity. If we have built our structure properly, then we can slow this exchange rate down but never eliminate it.

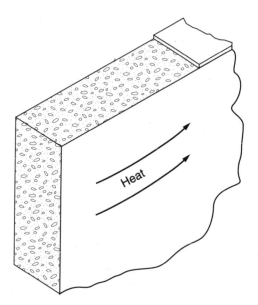

■ **3.4** *Heat moves along an isothermal plane looking for a path out.*

The American Society of Heating, Refrigerating, and Air Conditioning Engineers (ASHRAE) can provide accepted calculation methods for determining how and where heat energy will be transmitted through walls, especially solid or foundation walls. The method of calculation is called the *isothermal planes analysis method* or sometimes the *series/parallel method.* This method allows home builders and designers to predict the heat losses due to heat energy movement along isothermal planes.

Thermal Bridges

Thermal bridges are unwanted by-products in residential construction. Unfortunately, these "heat conduits" are created as the structure is being put together. A *thermal bridge* is the penetration of the insulation layer in a wall or assembly (doors, windows, etc.) by a noninsulating material. Wherever materials such as steel, concrete, aluminum, or other highly conductive materials are present, the possibility of a thermal bridge exists. Figure 3.5 shows a thermal bridge that is seen in many houses. Steel conducts energy hundreds of times faster than insulation, and concrete does not insulate. Energy loss attributed to the effect of steel and concrete thermal bridges can be higher than 70 percent.

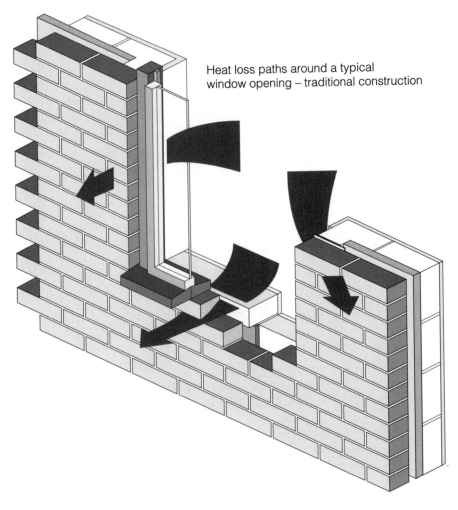

Heat loss paths around a typical window opening – traditional construction

■ **3.5** *Thermal bridging effect.*

The heat energy that has been created within a residence by its heating system begins to seek a path to exit the structure. A concrete wall behaves much like a radiator. The valuable heat energy moves along the isothermal planes parallel to the insulation. These are typically the reinforced concrete layers, and when the heat reaches a steel or concrete thermal bridge, the energy is then transmitted through the wall and out of the residence.

If a house could be built without windows, doors, and other openings, then energy savings would be more straightforward. However, these aforementioned "openings" have to be considered. They have to be considered for (1) their actual contributions to heat loss due to their material and makeup and (2) the losses around them due to the fact that we have just put a *hole* in the house. We will discuss in later chapters the heat transfer across windows and through doors, but as part of this discussion centering on structure and design, we need to briefly consider door and window openings. They have to be considered here because they do degrade the structural integrity of the residence somewhat. The more holes or openings that go into a shape, the less rigid the shape becomes. Optimizing the quantity, locations, sizes, and shapes of these openings is paramount. Often the doors and windows are viewed as "standard" amenities to house and room design, but they should be looked at a little closer, particularly because of the thermal bridging effects and how they can wreak havoc later in the house's life.

Thermal bridges typically are found at exterior walls, where the temperature difference between inside and outside is at its greatest. The cavities in these openings are either closed with no insulating materials, or the insulating materials used are inferior in insulating value. The nasty result is that heat will take a path of least resistance and try to move around the double-pane glass and through and across voids and cavities around the entire window or door opening. Besides a tremendous amount of lost heat, the resulting reaction produces condensation. With excessive condensation, plaster and paint deterioration at these locations becomes apparent. Mold can begin to grow while moisture stains appear, and the problem is now harder to deal with.

A certain amount of planning during the design stage and predicting where thermal bridges may be present can pay big dividends later. The first dividend results from not having to find them later, which can involve complicated heat-loss equipment, or even from not seeing their presence in the form of condensation and mildew. However, if thermal bridges are present, maybe in an existing house, then ASHRAE has acceptable methods for testing for their presence. One method is called the *guarded hot box device,* and it can detect and measure thermal bridges. Another agency that has performed extensive tests on thermal bridges is the Construction Technology Laboratory.

Dead Air Space

As important as it is to consider the best structural components for a house's foundation and framework, it is also important to plan some dead air space into the design. This concept may sound as though it is all right to waste construction dollars, but from an energy efficiency standpoint, this is not the case. Also, for the multitude of people who want *very* big homes, why not take this excessive size and make it work for you while it makes a statement? Dead air space placed in the proper locations will easily save thousands of energy dollars over the life of the home.

A true vacuum allows virtually no heat transfer at all. This would be the ideal solution for thermal bridges, but it is not practical. However, the next best "thermal break system" is the dead air space. As can be seen in Table 3.1, dead air space actually has an R value. The thickness of the dead air space, where it is in the building envelope, and whether or not it has a reflective surface will affect its resistance (R) value. Placement is important, but so is the "deadness" of the air in the space. If a draft or venting effect is present in the dead air space, then heat can still be lost through convection. Thus it is important to account for the right types and amounts of dead air space all around the residence.

Concrete Wall Construction and Thermal Mass Effects

Concrete is a structural material made up of aggregates such as sand and gravel (chemically inert substances) bonded together by water and cement. Since concrete may make up over 50 percent of the total building material weight in a house, it deserves attention. Additionally, since a house needs a solid foundation material, concrete needs even more attention given to it. How much concrete should be exposed? How thick should the walls be? How high do we make the walls? Should they be our living area walls as well? There are many questions, and they are justly asked because this structural mainstay is a critical element in a residence's total energy system. Understanding how heat moves through concrete will allow us to contain it.

To analyze this concrete "mix" and its elements from an energy basis, one finds that this low-cost, durable, and flexible building material is horrible as an insulator. An inch of polystyrene is nearly 10 times better as an insulator than concrete. However, polystyrene cannot absorb and retain heat like concrete does. Concrete comes in basically four configurations:

1. Tilt-up sections that are cast at the site and then lifted into place

2. Precast sections that are cast flat at the manufacturing plant and then transported to the site

3. Poured-in-place systems that use forms to allow for site casting in place

4. Prestressed sections that have been cast at the factory with any tensile components having some tension on them to give a prestressed condition as the pour is made

■ **Table 3.1 Typical Dead Air Space *R* Values (ASHRAE).**

Orientation and Thickness (in) of Air Space	Direction of Heat Flow	*R* Value for Air Space Facing		
		Non-reflective Surface	Fairly Reflective Surface	Highly Reflective Surface
Horizontal ¾	Up*	0.87	1.71	2.23
4		0.94	1.99	2.73
¾	Up+	0.76	1.63	2.26
4		0.80	1.87	2.75
¾	Down*	1.02	2.39	3.55
1½		1.14	3.21	5.74
4		1.23	4.02	8.94
¾	Down+	0.84	2.08	3.25
1½		0.93	2.76	5.24
4		0.99	3.38	8.03
45° slope ¾	Up*	0.94	2.02	2.78
4		0.96	2.13	3.00
¾	Up+	0.81	1.90	2.81
4		0.82	1.98	3.00
¾	Down*	1.02	2.40	3.57
4		1.08	2.75	4.41
¾	Down+	0.84	2.09	3.34
4		0.90	2.50	4.36
Vertical ¾	Across*	1.01	2.36	3.48
4		1.01	2.34	3.45
¾	Across	0.84	2.10	3.28
4		0.91	2.16	3.44

One side of the air space is a nonreflective surface.
*Winter conditions.
+Summer conditions.

Any concrete or concrete block foundation walls not covered by the surrounding earth are going to lose heat quickly. Even if the concrete has been absorbing heat all day long, thermal bridge effects will allow heat to move across the surface seeking the "shortest" path out. If steel or metal reinforcing is used, the heat transfer across these components is even more dramatic. This is the big reason why concrete has only been viewed as a foundation item and usually is covered as much as possible by the surrounding earth. Today, the thermal mass-absorbing properties of concrete need to be exploited.

Using the thermal mass properties of concrete-based materials, the structure can attain innate energy savings and also achieve an excellent building foundation as well. However, if not designed and implemented properly, a concrete or concrete block wall can be a source of tremendous heat loss. Thermal mass can be likened to the "heat inertia" of a material. The thermal mass effect, also known as the *M factor,* a little less is the concrete's ability to store energy. It also has a dampening effect with respect to temperature changes, especially for heating and air-conditioning systems. Materials such as concrete tend to remain cold whenever the ambient temperature is cold but can absorb heat over a period of time and retain it for a length of time. Thus orientation of the structure with respect to the ambient surroundings and to the sun is critical to concrete being used as a heat absorber. Likewise, annual weather conditions can play an important role in the structure's heat losses and gains. Yearly averages for cloudy days in a region also should be considered when seeking to use the concrete for thermal mass effects. If a particular region does not get enough sun-filled days during the year, then the premise for a thermal mass effect system is hard to justify. Any extra costs associated with this design will be difficult to recover if the "sun does not shine."

Thermal mass effects can be incorporated throughout the house. Interior walls can be made from stone, adobe, or brick. They can be oriented such that they can absorb heat from the sun during the day. Large fireplaces, several concrete slabs, and thicker floors of cement all help to retain heat via their thermal mass properties. The more heavy, heat-absorbing materials in a home, the better it is when it comes to building with thermal mass.

Consider the following example: A 30 × 40 ft house is on a 9-in concrete slab. The slab warms to 75°F from sunlight it absorbs through southern windows. Using an outdoor temperature of 25°F and an indoor temperature of 70°F, and given that the entire house loses 300 Btu/h, we can predict the total heat loss over a given period of time (such as overnight). The total heat loss is equal to the rate in British thermal units (Btu) per hour, 300, times the temperature difference, times a period of time (use 12 h in the example, i.e., 6:00 P.M. to 6:00 A.M.). Thus 300 × (70 − 25) × 12 = 162,000 Btu lost over that period. Table 3.2 shows different specific heats and heat capacities of some common materials. Using these data for concrete, we can see the effect of thermal mass properties of the concrete in action. With a total of 900 ft^3 (30 × 40 × 0.75 ft) and a heat capacity of nearly 32 Btu per cubic foot, the concrete slab can effectively store 28,800 Btu for a 1-degree rise in its temperature. For a 1-degree drop, it releases the same amount of British thermal units. If the slab drops 5 degrees from

■ Table 3.2 Specific Heats and Heat Capacities of Common Materials

Material	Specific Heat (Btu/lb/°F)	Density (lb/ft)	Heat Capacity (Btu/ft/°F)
Water (40°F)	1.00	62.5	62.5
Steel	0.12	489	58.7
Cast iron	0.12	450	54.0
Copper	0.092	556	51.2
Aluminum	0.214	171	36.6
Basalt	0.20	180	36.0
Marble	0.21	162	34.0
Concrete	0.22	144	29.0
Asphalt	0.22	132	29.0
Ice (32°F)	0.487	57.5	28.0
Glass	0.18	154	27.7
White oak	0.57	47	26.8
Brick	0.20	123	24.6
Limestone	0.217	103	22.4
Gypsum	0.26	78	20.3
Sand	0.191	94.6	18.1
White pine	0.67	27	18.1
White fir	0.65	27	17.6
Clay	0.22	63	13.9
Asbestos wool	0.20	36	7.2
Glass wool	0.157	3.25	0.51
Air (75°F)	0.24	0.075	0.018

75 to 70°F, it will release enough heat to virtually replace that lost by the house through the night.

Even though some assumptions were made and the example was not concerned with any other heat-loss factors (e.g., infiltration, etc.), this example gives you an idea of how much impact a concrete component or any other thermal mass component can have in helping to retain heat. Obviously, with today's controls and sensors, this thermal mass effect could be maximized, especially for summer and winter months. Automatic control of the actual gain during the season could be controlled adequately as part of a home automation system.

For years, concrete-based products had to have further work done to them to make the insulating values better. Often steel and other metal ties would have to be immersed in the concrete, and this had a more dramatic effect on energy savings—thermal bridging. Concrete itself acts as a thermal bridge when compared with other materials. A short circuit for our precious heat to escape was not desired with concrete-based products. Concrete-based products encompass cement or concrete-poured foundations, concrete block foundations, and other cement-containing materials. Brick, stone, and other masonry components are very similar in thermal mass effect but are not used so much any more for foundations but rather for decor and finish. All these materials have an ability to absorb and store heat on sunny days, and all in their normal state will allow heat to transfer quickly. Thus home builders have had to insulate and refinish these walls to stop heat loss.

As can be seen in Fig. 3.6, an 8-in-thick concrete wall made with sand aggregate has an *R*, or resistance, value of 0.64. Concrete block at 8-in is better, with an *R* value of 1.72, mainly due to the dead air space within. In either case, both values are not good when it comes to heat loss. However, add some insulation and sealant, and what a betterment in *R* value, as shown in Fig. 3.7. If one had some extra funds, the double foundation wall section shown in Fig. 3.8 would be ideal.

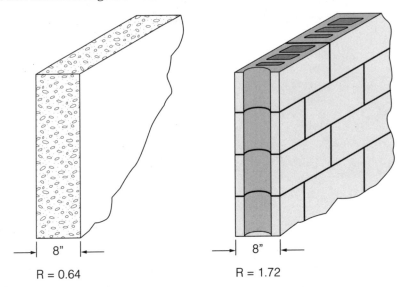

R = 0.64   R = 1.72

■ **3.6** *Thermal resistance differences between equivalent thicknesses of concrete and concrete block.*

Technology has brought us new concrete-based products that have built-in insulation. Whereas concrete and concrete block products were shown earlier to have poor insulating values, newer processes produce composites that give concrete a better *R* value. Some concrete has fibers of material mixed in, and these may improve its insulating value but possibly weaken it. Others incorporate an internal layer of insulation between wythes (Fig. 3.9). This technology gets the most out of the concrete; in terms of "thermal mass effect," the effect is still there, and by incorporating insulation materials into the mix as the concrete is mixed and poured, the concrete gets insulation qualities as well.

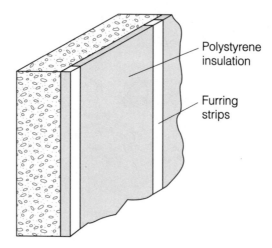

■ 3.7 *Addition of insulation to a concrete wall provides an increased thermal resistance.*

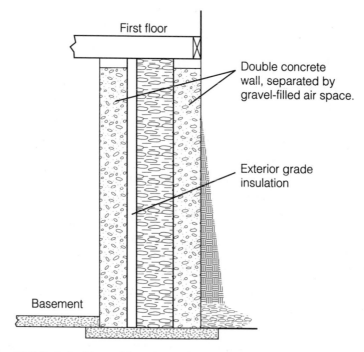

■ 3.8 *A double-wall foundation with insulation on an outside wall.*

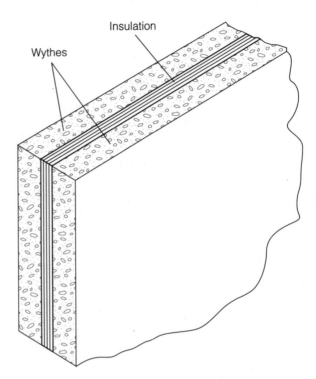

Insulation

Wythes

■ **3.9** *A layer of insulation between wythes of a concrete wall.*

R values for concrete systems that use insulation board depend on the final thickness of the wall, the type and thickness of the insulating board, and any other modifications made to the wall. Obviously, by introducing insulating board into the composite, the *R* value will rise dramatically. Additionally, no thermal bridges are present in such a system.

Thermal mass systems should be present in every home. They are of low cost, many times decorative, and quiet contributors to the energy-savings bottom line. Think in bulk when trying to achieve energy savings in new home construction.

Steel Versus Wood Construction

Wood framing has been the standard in residential construction for hundreds of years. Early on, it was the only way to frame, and it is still employed today. Wood construction has both advantages and disadvantages. As more and more energy systems become less dependent on wood's inherent insulating effect, steel should evolve as an excellent framing choice. The ecological movement to save trees and forests helps to solidify the argument for steel framing, but not until methods are developed to insulate and stop all thermal bridges associated with steel studding (remember, steel is not friendly to the energy-conscious home owner because it short circuits heat well).

Wood frame construction, although the mainstay for contractors, has lost some of its position in recent years. Fluctuating wood prices, poorer quality in the wood itself, and newer alternative technologies have added to the environmental concerns over building with wood. And while light-gauge steel is gaining popularity as a framing choice, the environmental arguments continue. The arguments change depending on who is presenting the argument (the American Iron and Steel Institute, or AISI, will promote what?). It should be mentioned, since this is a book about energy systems, that it *does* take energy to make steel and form steel structural members, and there is residual waste—all things that industrial processes contribute to the environmental issue. However, this process can produce consistent components, since warping and twisting typically are not seen in finished steel studs, as sometimes happens with wood studs. Other defects are also minimized in steel members. Steel framing can, in fact, be substituted rather easily by most contractors because the prices are consistent, steel studs can easily replace wood, and steel joists and beams are attractive over wooden built-up beams and wood floor joists.

From an energy-systems vantage point, and considering structural needs for any residence, steel framing can be an efficient choice. Its biggest drawback is that its thermal efficiency is poor, but this can be dealt with if thermal efficiency is designed into the structure. Table 3.3 shows equivalent

■ **Table 3.3 Wall *R* Values for Typical Configurations of Steel or Wood Framing (ASHRAE)**

Framing Member and Spacing	Nominal Cavity Insulation	Combined Cavity and Framing *R* Value (Without Sheathing or Air Films)	
		Wood Framed	**Steel Framed**
2 × 4 16 in o.c.	*R*-11	*R*-9.0	*R*-5.5
	R-13	*R*-10.1	*R*-6.0
	R-15	*R*-11.2	*R*-6.4
2 × 4 24 in o.c.	*R*-11	*R*-9.4	*R*-6.6
	R-13	*R*-10.7	*R*-7.2
	R-15	*R*-11.9	*R*-7.8
2 × 6 16 in o.c.	*R*-19	*R*-15.1	*R*-7.1
	R-21	*R*-16.2	*R*-7.4
2 × 6 24 in o.c.	*R*-19	*R*-16.0	*R*-8.6
	R-21	*R*-17.2	*R*-9.0

resistances for various sized steel- and wood-framed walls. This table uses the parallel-path method for calculating, and as can be seen, the steel-framed walls never really provide any substantial builtup in R value. Throw in a few nasty thermal bridges, and steel becomes extremely unattractive. However, finishes over the inner and outer surfaces of the walls can make a tremendous difference. Exterior spray systems, such as Dryvit, can provide both an obstacle to thermal bridging and added R value. They also can provide a decorative, stucco-like finish. Cost savings associated with the use of steel studs (lighter weight, easy to handle, less labor, wire and plumbing holes already predrilled into the studs, and base price of light-gauge steel studs) allow for other measures to control heat loss (even if some extra costs are incurred).

If measures have to be taken around door and window openings anyway to eliminate thermal bridging and an exterior finish other than brick or siding is to be employed, then the choice of steel as a structural component is sound. Steel is inert, so indoor air quality is much better and safer in case of fire. Wood releases terpenes, and treated wood contains toxins. Additionally, steel does not burn but can weaken in a fire, so this chapter cannot properly address this issue. As far as thermal issues related to steel construction go, there are other options.

One option to better the thermal bridging problem associated with steel studs is to modify them, not after they are in place but before. Manufacturing different configurations is one approach. As can be seen in Figs. 3.10 and 3.11, two configurations are possible. Figure 3.10 shows a stud with numerous holes. The intent of these holes is to provide numerous breaks in the thermal bridge so as to slow down the heat migration through the steel. Figure 3.11 shows another configuration aimed at reducing the thermal transfer rate. Here several protrusions are introduced to each flange of the stud. The intent here is to minimize actual stud to other wall component contact. A small amount of dead air space acts as an insulator. Of course, the short-circuit path for the heat to take is not eliminated altogether, and thus losses will still occur. The extra work and cost to attach a thermal break or barrier on one side or the other of the steel stud are still necessary. This is the most common option employed, and insulating board and insulating sheathing are commonly used and will bring the R value up. There are several finishing systems available at this point in the construction phase. Many will be discussed in Chap. 7.

Another option is to depart from conventional platform framing and develop new framing methods. Steel, by virtue of its uniform strength, can span greater distances than equivalent wood studs (gauge of steel versus standard 2×4's). Thus creative house plans can be achieved and different thermal barriers can be employed (such as foils, insulations, etc.). It is important to note that any interior wall of a house could be framed in steel because no heat transfer could occur out of the house from these points.

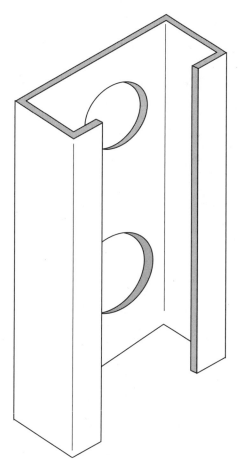

■ **3.10** *A metal stud with holes to provide various breaks in the thermal bridge.*

A wall with wood studs, 16 in on center, will have a high proportion of its given surface area apportioned to heat loss at the R value of 3.5 in of wood. This is good, and if the entire wall were solid wood, the insulating value would be excellent. However, this is not done in conventionally framed homes, but the log cabin does incorporate this concept into its design. Wood framing has been the standard for so long that any departure from it in mass will be slow in coming. Lumber that is air dried is a better selection for wood studs because it has a lower content of latent energy.

Whenever exterior walls are framed with steel studs, place the studs on 24-in centers. A heavier gauge may have to be used for a more solid wall, but the fewer steel studs means less thermal bridging. Moreover, extra cavity insulation may not be worth the investment because any advantage gained as a result of thicker walls or higher-density insulation is basically canceled

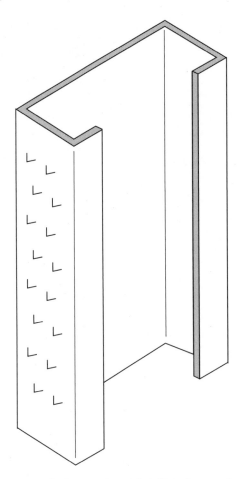

■ **3.11** *A metal stud with nibs and small protrusions on the flanges to reduce the thermal transfer rate.*

by heat that is lost via short circuit through the steel studs. This money is better used to place insulating sheathing over the studs as a thermal break and a full, continuous layer of insulation. Last, try to limit the amount of fasteners directly to the studs. They also act as thermal bridges. Instead, install plywood, sheathing (OSB), or wood straps on the studs to facilitate attachment of any insulation and to minimize the contribution of metallic fasteners to the thermal bridge effect.

Steel framework still has a long way to go before it ever completely displaces wood as the choice framework element. The thermal reasons are the most dramatic, and when they are resolved suitably, steel studding will become an easy choice. We may get forced into an all steel structural system as the norm whenever the environmental concerns reach crisis proportions. However, until then, research and development continue at a steady pace, and the pricing is still not that bad—yet!

Geodesic Domes and Round Houses

These structural systems are not for everyone. While the energy savings due to the shape and use of materials in these structures are exceptional, the aesthetics and appeal of the "round house" are not for most. The concept of round houses and geodesic domes is not new. The traditional "igloo" is a testimonial to the fact that the shape is well suited to severe and colder climates. Geodesic domes are very similar in design, but with conventional building products. These conventional building products and the use of control systems can only better the potential in energy savings with these types of structures.

A round house should be differentiated from a geodesic dome, and vice versa. Each has one thing in common—neither is actually round. Conventional building components are square, and thus triangular and geode shapes will be used to construct these types of homes. They can have eight, ten, twelve, and even more sides. The floor plan of a round house is shown in Fig. 3.12. Note that the plan clearly shows that there are many

■ **3.12** *A geodesic dome floor plan.*

physical sides to the structure—it is not round. Where round houses and geodesic domes differ is in their respective elevations. Round houses still contain upright walls onto which cold winds will pound, whereas domes become more spherical in shape. This spherical shape causes winds to glance off the surfaces. Round houses still require a separately constructed roof system, and domes do not. The dome's roof is actually its walls; they blend together. This helps dramatically in heat-loss calculations and in actual losses.

The geodesic dome is a product of several smaller components such as triangular and polygonal shapes that fit together and work together. The shapes are all connected together to form a unique building system. The base shape, as seen in Fig. 3.13, is triangular at the start, and base shapes get fitted to other adjacent members to form a strong polygonal shape. Thus this pattern building makes the dome a spherical structure capable of being set directly on the ground as a complete structure. The multiple shapes, when placed together as a system, use tensile properties to replace the arch principle and distribute stresses throughout the dome, much like an egg does. For its size and shape, an egg is very hard to crush when pressure is applied on the ends toward each other. The concept and design of the geodesic dome were developed by R. Buckminster Fuller in the 1940s and 1950s. Since then, thousands of these structures have been built around the world.

■ **3.13** *The base shape from which geodesic domes are made.*

The geodesic dome is made up of many energy systems all working together. The dome's shape, surface area versus actual livable square footage, and other pronounced features due to its design make it a necessary element when discussing energy. In fact, most people who venture to take on this design have energy conservation in mind every step of the way. Heat loss in typical uninsulated houses is broken down in the following percentages:

Roof, 25 percent
Walls, 25 percent
Windows and doors, 25 percent
Floors, 5 percent
Infiltration, 20 percent

Since a dome's walls are its roof and vice versa, there is less physical area from which to lose valuable heat. If we take away 25 percent heat loss due to roofs (since we now have a wall acting as the roof) and reduce infiltration by 10 percent (since the winds skim over the rounded top of this structure), we have saved 35 percent of the total heat loss and have not done anything else to conserve the heat. Of course, this example assumes a perfect scenario, and the results are probably exaggerated. However, the point is well made: The shape of the structure can yield energy system benefits immediately!

A geodesic dome typically is not constructed as are conventionally framed homes. It is similar in that a foundation and subfloor system (Fig. 3.14)

■ **3.14** *Subfloor and foundation system for geodesic dome construction.*

have to be in place to set the multiple geodesic members onto. This phase also involves planning and attention to detail, since this will set the stage for further energy savings. As is seen in Fig. 3.15, the construction actually involves attaching multiple triangular members together to form a much stronger polygonal shape. In turn, these polygonal shapes create a larger, stronger spherical shape that becomes the shell of the geodesic dome. The building of the dome shell (Fig. 3.16) and the final geodesic dome shell (Fig. 3.17) are the base elements of the structure. Sealing of openings, minimizing thermal bridges, waterproofing, and full insulation of the members have to be accomplished in order to achieve maximum energy savings with this type of structure.

■ **3.15** *Assembly of triangular shapes to make a polygonal structure.*

The polygonal members can be built from 2 × 6 in framing lumber, and thus 6 in of fiberglass insulation can be used for a good *R* value. Even 2 × 4 in frames, when insulated, will carry a good *R* value (obviously not as high as 6-in frames). Since the walls and roof are built up to be one uniform spherical structure out of the same material (typically wood), thermal bridges are not an issue until door and window openings are addressed. These openings and the possibility of thermal bridges have to be finished in the same manner as a conventional house finishes them so as to eliminate the heat-transfer short circuit.

All in all, a geodesic dome is a very sound structure. Structurally, its internal tensile strength is tremendous. Its shape lends itself well to climates

■ **3.16** *Building of the geodesic dome shell.*

■ **3.17** *Final geodesic dome shell, ready to finish.*

with harsher weather. It can be insulated and sealed as well as its corresponding conventionally framed counterpart in the residential world. It does have many energy-saving advantages inherent to its design, but again, it is not for everyone. However, it should be evaluated whenever embarking on a new home. The exercise in comparisons between conventional square homes, custom homes, and these geodesic domes will be worth the time.

Designing the Passive Solar System into the Structure

Taking advantage of modern technology in designing energy systems, passive solar systems can be optimized. They actually can be renamed *electronic passive solar systems* (EPSS). Dark heat-absorbing colors, thermal mass effect materials, orientation and shading with respect to the sun, optimum sizing, and the residence's shape are just a few of the answers to the who-what-when-where-why questions. However, as electronic controls become more cost-effective, prevalent, and available, their use will easily exploit solar energy systems.

For years, every home owner has had his or her idea of what degree of energy savings they wanted and needed in their residence. They made "energy sacrifices" as bigger, more showy facilities presented themselves and won out. However, most persevered with different attempts to save residential energy. Some ideas actually have become accepted standards or guidelines. In winter, the house should absorb solar heat by day and retain it at night. In summer, the house should reject heat during the day and release it at night.

The thermal mass effect has evolved into a big player in passive designs. So have many other building system variations. Venetian blinds, shutters, fans, pumps, and other motorized devices can be better controlled to optimize any heat gain. Passive solar systems employ every method possible in trying to soak up the sun's rays. The key is to retain the heat once it has been "snagged," i.e., build the house as a heat "trap." Likewise, heat from the sun may not be required during the hot summer months, and therefore, other measures may have to be taken to limit the amount of heat retained by the thermal mass components. Shading becomes a very important part of controlling the heat gains during this phase. Many of these solar and sun-related topics are covered in greater detail in Chap. 11.

As mentioned earlier, color selection for surfaces and their absorption rates also can have a dramatic effect on a structure's full capability to retain heat from the sun. Sunlight striking a surface will either be absorbed or reflected. A more in-depth look at the absorption and emissivity of materials is provided in Chap. 11. *Emissivity* is the amount of thermal radiation emitted from a material divided by the radiation from a black body, which is assumed to have an emittance value of 1. Color does play an important role in how much extra heat we desire to be absorbed and where in the residence this should occur.

By orienting the structure appropriately with respect to the sun, and by selecting materials and colors for those materials judiciously, we can optimize our heat gain position. Not only do we get the thermal mass effect working for us but we also get enhanced performance on the color coating given to the thermal mass component. Table 3.4 shows various materials and their corresponding absorption values. For example, black tar paper has an absorption rate of 0.93, which means that it can absorb 93 percent of the incoming sunlight, whereas aluminum foil has an absorption value of 0.15, so most of the sun's energy would be reflected by this material. Thus careful choices in building materials for the structure and, more important, their colors will have a profound effect on our residence's ability to retain valuable sun-generated energy.

■ Table 3.4 Absorption Properties of Colored Materials

Material	Absorption Value
Aluminum foil	0.15
Black tar paper	0.93
Concrete	0.60
Dry sand	0.82
Flat black paint	0.96
Fresh snow	0.13
Galvanized steel	0.65
Granite	0.55
Graphite	0.78
Green paint	0.50
Green roll roofing	0.88
Gray paint	0.75
Red brick	0.55
Red paint	0.74
Water	0.94
White enamel	0.35
White paint	0.20
White plaster	0.07

Note: An absorption value of 1.00 represents 100 percent absorption.

Closing Remarks on Structural Energy Systems

R. Buckminster Fuller once wrote:

> . . . I did not set out to design a house that hung from a pole, or to manufacture a new type of automobile, invent a new system of map projection, develop geodesic domes or Energetic-Synergetic geometry. I started with the universe—as an organization of energy systems of which all our experiences and possible experiences are only local instances. I could have ended up with a pair of flying slippers.

This paragraph sums it up: The universe *is* an organization of energy systems. Every facet of building construction and every chapter in this book are dedicated to these energy systems found within. They all interact and together can provide us with comfort and efficiency.

No matter what other energy-saving measures a home builder or home owner is planning, they must put emphasis on the structure. Their energy plans could be hindered or helped by the framework and structure. Once the foundation is in place and the framing is up, it is very difficult and costly to make corrections—it's too late. The up-front planning or modeling is important. Seek to use a builder who can provide graphic pictures or diagrams of concepts and designs. A residence is actually a complicated system, complicated in the sense that many factors influence its design: ergonomics, finances, comfort, needs, standards, codes, and obviously, energy efficiency. It is not just a place to "hang one's hat." Thus the builder should be capable of working with the client on each concept. Being able to model a concept, show it graphically, and discuss it in detail will help to eliminate the bad designs and focus on the most appropriate ones.

Modern Residential Building Construction

Building construction has evolved over the years, yet the basic elements of the shelter have remained the same. Protection from hot and cold climates, wind, rain, and snow, is still the often overlooked premise for the residential structure. Today's home is more of a statement, many times departing from energy conservation and sometimes blatantly disregarding energy savings at all. This may be a reflection of a strong economy and a wholesale societal change for the majority; however, energy efficiency should be built into any residence, particularly for the long haul. If a house is to last for 100 years, then the rising costs of fuels and energy have to be considered because the house will see multiple owners with varying needs over the course of its existence.

Earlier in this book we looked at the structural and primary building techniques designed to gain energy efficiency. Now we must look at the secondary components and material systems necessary to keep saving energy.

Standard, Conventional Design, or Built for Energy Savings?

For years, residential buildings have been built in much the same manner. It is sometimes tough for builders to take drastic departures from accepted standards. They can get complacent with tried and true methods for safe, fairly energy sound structures. To realize better energy efficiencies, however, especially in the future, home builders are going to have to try new materials and methods and basically experiment. This will have to be the case if higher-efficiency residences are to result.

Conventional, Western-style platform framing for multiple-story housing has been the norm for decades. Not until recent years have codes and agencies set higher energy-efficiency standards with which all new homes must comply. These requirements will only get more stringent, especially as either energy sources get scarce or prices go out of sight (possibly both can happen simultaneously). With electronics, software, and home automation products also demanding their share of the energy-efficient home of the twenty-first century, home builders are going to have to change with the times and depart from the conventional. Homes will become a conglomerate of energy subsystems all working somehow in unison to provide the very best energy usage possible.

Doors

A door is actually a door system. The door is a type of mechanical system that has a major effect on a residence's heat loss. The door assembly, especially an exterior door, is made up of hinges, a lockset, sometimes a closure, a threshold, a full frame, insulation, some weatherstripping, and the door itself. From an energy standpoint, it represents a large hole in the wall with the capability of introducing a lot of cold winter wind and a convenient path for heat to escape. Granted, doors are necessary for egress and ingress of the human occupants, and they should have a good appearance. However, they have various cracks around their opening and they have to be addressed in a special way whenever a heat-loss calculation is done for the house.

A door is a barrier of wood, metal, glass, or a combination of materials that is used to fill the void of a rough opening in a wall. These openings can vary from approximately 3 ft wide by 7 ft high for single doors to 6 ft wide by 7 ft high for double doors. These represent holes in the wall of 21 and 42 ft^2, respectively. A lot of air can flow through such openings, and a lot does whenever a door is left open for any length of time during the winter months. Fortunately, the door system works together to close the opening and minimize the heat losses around that opening.

With exterior door systems of today, the door frames, doors, and door hardware all work to keep heat loss down. The door frames have thermal breaks of air spaces and insulating materials to stop the conduction of heat, whereas the thresholds and weatherstripping work to reduce air infiltration and exfiltration. The doors themselves are also equipped with generous amounts of insulation and their own thermal breaks. There are also many types of doors.

Solid wood or hollow metal doors with insulation offer the best energy efficiency and are used most often. However, many doors are equipped with glazing, which has to be treated much like a standard double-pane window in terms of heat loss. The sliding glass and atrium-type swinging double doors usually contain a high quantity of glazing versus wood or vinyl cladding and thus lose more heat than do solid doors. R values for typical doors are addressed in Chap. 6. Louvered doors offer some ventilation but have limited use.

As long as the threshold height is adjusted properly (Fig. 4.1), the weather-stripping seals tightly against the door whenever it closes, and the frame is well caulked, the door system will suffice very well in the opening of the house. Shim areas around the door should receive expanding-foam insulation, whereas other cracks should be caulked. To measure whether or not a good seal has been made around a door, take a credit card and try to slide it between the seal/weatherstripping and the door. There should be some actual resistance to the card's sliding in if a good seal is made between the door and the weatherstripping. If there appear to be gaps, then you may have to adjust the weatherstripping somewhat to get the seal. Also check around the lock set area for air leakage, since this area often allows heat to transfer from the house to the outdoors. If this is the case, further adjust the weatherstripping at that point or glue more weatherstripping material over the top of that which is there. However, it pays to treat doors with keen interest when making a selection because they differ in construction and in their thermal resistance. Door size, color, and thickness are all critical elements to their energy-efficiency contribution to the house. Weigh these factors as well.

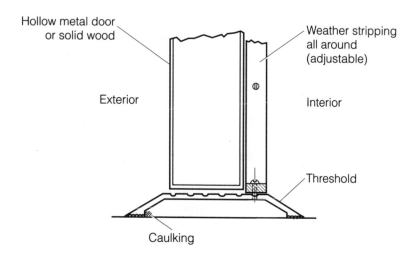

■ **4.1** *An exterior door with threshold and weatherstripping.*

Windows

Windows are the most prominent architectural aspect of residential buildings. They serve a variety of purposes. They make the residence look nice, they provide a view of surroundings; they let in daylight for natural lighting, they provide for natural ventilation, and they can be used as emergency exits. It is often forgotten that windows have one of the greatest energy impacts of any residential building element. They can be major sources of heat loss if not given proper attention. The heat losses and gains of a building occur by the conduction of heat through the solid elements of the building envelope (i.e., walls, ceiling, floor, windows, and doors). This heat is then convected and radiated to the outside or inside depending on the direction of heat flow. Air infiltration (air leaking in) and air exfiltration (air leaking out) are also important contributors to a building's heat loss or gain. Solar heat gain can offset winter heat losses but contributes to summer cooling loads.

Solar radiation is the most efficient source of heat that can be used to warm a house. However, solar gains are also the source of unwanted heat, especially in the summer. Because windows transmit much more solar radiation than the opaque elements of a building, they are the primary focus for controlling the timing and amount of solar gain. The orientation of a window greatly affects the times and levels of solar heat gain through the window. South-facing windows (for those residences in the northern hemisphere) see the sun throughout much of the day. East- and west-facing windows face the sun primarily in the early morning and late afternoon hours, respectively. North-facing windows view the sun only obliquely during the cooling season and not at all during the heating season.

The airtightness of a window depends on the sash/frame construction and fit and installation in the wall. Infiltration, or air leakage between the indoors and outdoors through cracks and joints around window frames, sash, and glazing panes, makes additional heating and cooling necessary. Excessive humidity is often a problem in residential buildings because of the many sources of moisture. This is especially the case during the heating season when windows and doors are typically closed to conserve heat. Under such conditions, moisture condensation often occurs on window surfaces that are below the dew-point temperature of the surrounding humid air. In addition to poor visibility through windows, severe and repeated condensation also may result in damage to window frames and sills, paint and wallpaper, carpeting, plasterboard, and structural framing in walls.

In comparison with opaque building elements, windows have a disproportionate effect on the heating and cooling requirements of a home. Solar transmission through windows can account for 33 percent of the cooling load, whereas ordinary interior shading devices reduce this load to approximately 20 percent. Solar heat gains during the heating season have the reverse effect and are beneficial in reducing heating bills 10 to 30 percent. Residences designed to optimize passive solar performance can use even more solar gain. Infiltration around windows may account for the loss of 10 percent of the heated or cooled air in a home. Natural ventilation through windows can, in some instances, substantially decrease or even eliminate the need for mechanical cooling.

Some window types include

The *bay*, which is an angled combination of windows that projects out from the wall of the home (Fig. 4.2)

■ **4.2** *A bay window.*

The *bow*, which is similar to the bay window, but the combination of windows has a more circular, arced appearance (Fig. 4.3)

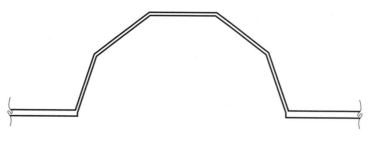

■ **4.3** *A bow window.*

The *casement window*, which cranks out, hinged on the side (Fig. 4.4)

The *awning-type window*, which is a top-hinged window that swings outward for ventilation (Fig. 4.5)

The *double-hung window*, which has an operable top and bottom sash (Fig. 4.6)

The *oriel window*, which is either a single-hung or double-hung window with the meeting rail not located in the center of the frame (Most oriel windows have a 60/40 split to their configuration.)

The *picture window*, which is basically a window that has no movable sash

The *slider*, which is a window with its sash moving horizontally instead of vertically

One type of window that is excellent protection against solar radiation is the low-E or argon-filled glass window. Low-E glass helps keep heat out in the summer and heat in during winter. Low-E glass is treated with thin transparent coatings of metal oxide and silver for improved thermal performance. It reduces the penetration of ultraviolet rays to minimize fading of carpet and draperies and blocks the indoor heat from escaping in the

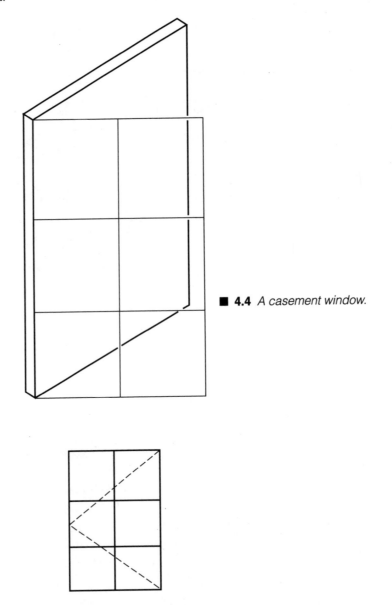

■ **4.4** *A casement window.*

winter months. During the summer, it reflects nearly 85 percent of the sun's infrared and ultraviolet rays that fade objects. At the same time, it allows visible sunlight to enter the room. Argon gas replaces the air inside the glass unit. Since it is heavier than air and is not in continuous motion like air, argon greatly reduces the transference of heat and cold by convection currents. Argon is an odorless, colorless, tasteless, nontoxic gas that is six times more dense than air. When replacing air between the glass panes to reduce temperature transfer, argon gas provides extra insulation.

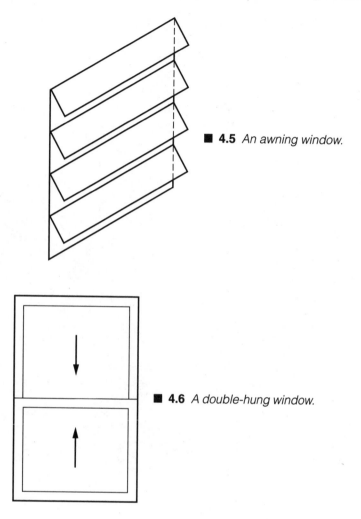

■ **4.5** *An awning window.*

■ **4.6** *A double-hung window.*

Moisture and condensation are both negative issues when it comes to windows. The *condensation resistance factor* is a measure of the effectiveness of a window or glazing system to reduce the potential for condensation. The higher the condensation resistance factor, the more efficient is the window or glazing system. An incorporated component to the window system is the *desiccated matrix*. This material is often used in insulating glass to absorb water vapor that causes fogging. Windows truly are more than a piece of glass covering an opening in a house. They are full, complete window systems with multiple components working together to provide as good an energy-efficient package as possible.

The installation of windows is also a process unto itself. The window frame and rough opening have to be fully square in order for the window to operate properly. This means the introduction of wood shims around the window frame. This can be seen in Fig. 4.7, and either foam insulation, caulking, or both will be needed to fill these large cracks to eliminate infiltration. As can also be seen in Fig. 4.7, there is a substantial amount of rough framing around the window opening so that there should be little movement of the members over time. Framing members will move and even swell with moisture, so it is very important to make this installation as sound and trouble-free as possible for future changes around the window. This movement sometimes causes windows to jam or to be hard to open and close.

■ **4.7** *Framing members around window.* (Courtesy of Wayne J. Henchar Custom Homes.)

Insulation

Today's residential building construction requires a certain amount of insulation in the walls, floors, and ceilings. This extra material provides a thermal resistance to heat in those areas of the house where it is placed. Thermal insulation decreases the flow of heat from a hot region to a cooler one, from the interior of the house to the exterior in the winter, and from the exterior to the interior during the summer. By conducting heat poorly, insulation probably has become the single biggest contributor to energy efficiency in the past three decades.

There is no perfect insulator, but a thin layer of nonmoving air is several thousand times better at resisting the flow of heat than is a good metallic conductor of equivalent thickness. This is illustrated in Fig. 4.8. This concept has resulted in the development of insulation of many types from

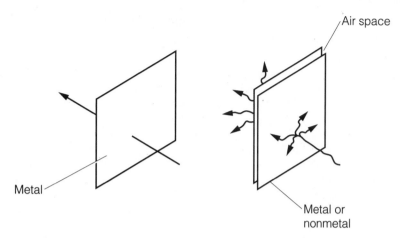

■ **4.8** *Nonmoving air within an air space allows much less heat through than an equivalent thickness metallic conductor.*

these two precepts: (1) that nonmetallic materials are better to use and (2) that tiny air spaces have to be either impregnated into or somehow used in conjunction with the nonmetallic material. The size of the air pockets has to be small; otherwise, any movement of air would produce convection currents that could result in the transport of the heat throughout the material (and eventually out of the space). An air layer of ¼ in, or 0.6 cm, is about the limit for maximum thickness before air movement occurs.

The increase in thermal resistance, or R value, when using insulation is dramatic. An example of the effect of insulation added to an exterior wall is shown in Chap. 6. However, Table 4.1 shows many of the different types of insulation used today, with their corresponding R values. Insulators that combine nonmetallic materials with air spaces include cork, fiberglass, magnesium carbonate, felt, rock wool, foamed plastics, and other elements. Other physical building materials, such as hollow glass bricks and concrete blocks, are good examples of materials that integrate small air pockets within (Fig. 4.9). Lighter aggregate blocks and even homogeneous concrete can be made into better insulators with the impregnation of air and less dense composition. Also, thermal pane windows use the air space as an insulator, and the thermal breaks in many other assemblies (frames, sashes, etc.) sometimes use a small air gap. This helps yield a resistance value of R-1.6 for windows and glass doors.

In conventional, modern-day residential building construction, typical amounts of insulation are recommended and used. The colder the climate, the more insulation should be used. In common stud walls, 3½ in of standard batt fiberglass insulation provides an R value of approximately 11 and should go as high as R-16 with new home construction whenever possible, whereas in the ceilings the recommended range is from R-19 to R-32.

■ Table 4.1 Resistance Values of Insulating Materials

Insulating Materials	Resistance *R* per Inch of Thickness
Blanket and batt	
Mineral wool, fibrous processed from rock, slag, or glass	3.12
Wood fiber	4.00
Cellular glass	2.44
Corkboard	3.57
Glass fiber	3.85
Expanded rubber (rigid)	4.55
Expanded polyurethane (*R*-11 blown)	5.56
(thickness 1 in and greater)	
Expanded polystyrene, extruded	3.85
Expanded polystyrene, molded beads	3.57
Mineral wool with resin binder	3.45
Mineral fiberboard, wet felted	
Core or roof insulation	2.94
Acoustical tile	2.86
Mineral fiberboard, wet molded	2.38
Wood or can fiberboard	2.38
Perlite (expanded)	2.63
Vermiculite (expanded)	2.08
Preformed roof insulation	2.78

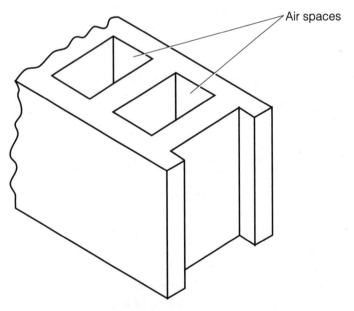

■ **4.9** *Integrating air pockets within a concrete block.*

Insulation can and should be used in other areas of the house, such as the hot water tank, hot water pipes, and ductwork that runs along outside walls (Fig. 4.10). By insulating behind the duct rather than only over it, warmed or cooled air is insulated from the outdoors, and many times insulation is used in rooms or areas of the house that need to be soundproofed from other rooms. Another advantage of using insulation is that some types allow for sound deadening (fiberglass and any type with porous, sound-absorbing features). However, a disadvantage is seen when the insulator gets wet. Any time that moisture is introduced to the insulator, the performance is reduced because the heat conduction of water at atmospheric temperature is more than 20 times that of air; therefore, this is something to avoid by making sure the insulation does not accumulate any moisture. Other areas within the house that can receive insulation and some recommended resistance ratings are as follows:

Exterior doors	*R*-4
Floors over unheated areas	*R*-19
Basement walls	*R*-10
Heating ducts in unheated areas	*R*-7
Slabs on grade	*R*-7

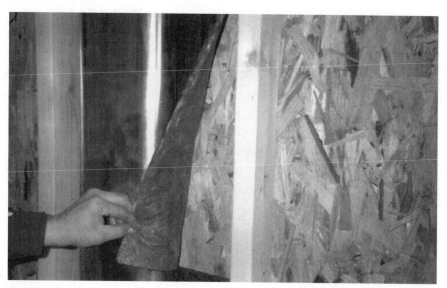

■ **4.10** *Insulating behind ductwork on exterior walls.* (Courtesy of Wayne J. Henchar Custom Homes.)

Since it is evident that insulation is a tremendous tool in making a residence energy efficient, there continue to be advancements in its development. Superthermal insulators were first developed by NASA for use in space, where temperatures could approach absolute zero, which is –459°F (–273°C). These composites consist of multiple sheets, each nearly 0.002 in thick, of aluminized Mylar. The Mylar is separated by thin air spaces at an equivalent 50 to 100 layers per inch of thickness. This technology is carrying over into the residential building construction marketplace.

Energy-Saving Tips and Maintenance Suggestions

The following energy-saving techniques, tips, and maintenance suggestions are provided for the home builder and the home owner. These apply to a multitude of appliances and energy-consuming products. They are meant as a guide for energy-efficiency-type ideas. Consult your local electric company and the businesses from which you buy your power and appliances, respectively. They often will furnish similar energy-saving tips and also can provide manufacturers' suggested maintenance schedules and recommendations. By practicing these sound energy-saving techniques (some are just common-sense items), you can decrease the amount of energy used in your home every month.

Many of the following suggestions pertain to cutting costs on monthly electric bills. Overall energy consumption in other areas also can be achieved by performing some of these various tips, including

Keep your monthly electrical bills, and monitor your usage each month. Look for seasonal trends and peak operating surcharges.

Limit the use of large appliances during any peak, higher-priced periods.

Shift usage of large appliances to off-peak, lower-priced periods.

Turn your air conditioner off when you leave home for 4 hours or more.

Whenever you are at home, set air conditioner at 78°F or higher. During the higher-priced period, it is recommended to increase temperature by 1 to 2 degrees of your normal setting.

Use fans to increase your comfort at higher thermostat settings.

Turn fans and lights off as you leave a room.

Wash clothing during lower-priced periods, if available in your electric district.

Have the air conditioner serviced once a year. Clean or change the air conditioning filter monthly.

Energy-efficient air conditioner maintenance should include following the manufacturer's routine maintenance directions, which may involve having the entire system checked once a year by a qualified service person. An annual service check should include

Lubricating the condenser fan bearings, especially if they are not permanently sealed.

Cleaning the air intake side of the condenser coil

Tightening electrical connections

Checking operating current with a volt-amp meter

Checking supply voltage with a volt-amp meter

Checking refrigerant levels and pressures

Inspecting the compressor start and run capacitors along with the motor.

Inspection of a furnace or air handler should include

Cleaning or replacing the filter

Lubricating an unsealed blower fan motor and blower bearings

Checking and adjusting the fan belt tension/condition

Cleaning the blower wheel

Inspecting and adjusting the burner

Cleaning the evaporator coil and checking fin condition

Cleaning and inspecting drain lines and pan

Checking for excessive attic air leakage into return chamber

Checking the thermostat for proper operation and calibration

Measuring air temperature at supply and return vents

Permanent filters should be cleaned with mild detergent, per manufacturer's recommendations, every 30 to 60 days. Replaceable filters should be replaced every 30 to 60 days.

Check for air leaks, particularly at the furnace or air handler. Seal leaks with duct tape. Ductwork should be insulated with at least 1½ in of batt insulation with a foil or vinyl vapor barrier on the outside. Tape up any leaky ductwork.

Keep doors and windows closed when the air conditioning system is running.

Caulk and weatherstrip to close air gaps.

Open drapes and shades over windows facing sun during the winter and close them during the summer.

Wash or dry clothes during warmest times of day, either late morning and afternoon.

Use the exhaust fan when cooking only when needed.

Keep the thermostat at the desired setting. Do not keep changing the setting.

The location of your refrigerator or freezer is important. Do not place it in direct sunlight or near any source of warm air, such as the range, dishwasher, or heating ducts.

What's inside counts, too. Refrigerators work more efficiently when food is arranged to allow air to circulate. Freezers, however, are more efficient when they are full.

Avoid opening the refrigerator and freezer doors frequently. Plan ahead. Take as many things out at one time as you will need.

Cleaning is important. Every 3 months, clean dust off the condenser, coils, fins, and evaporator pan. And refer to your owner's manual for other cleaning maintenance.

If you have a "power saver" switch, turn it on during humid weather to prevent moisture buildup. Then turn the switch off when the humidity is lower.

If you have a manual defrost refrigerator, do not let more than ¼ in of ice build up. Efficiency is reduced as ice buildup occurs.

Refrigerator and freezer door gaskets must be tight. Worn, dirty, cracked, or poorly fitting gaskets should be replaced to keep top efficiency.

When baking in your oven, do not be a "peeker." Twenty percent of oven heat can be lost every time the door is opened.

Try to use all the space in your oven. Extra dishes cooked at the same temperature can be frozen for future meals. (Cook it. Freeze it. Microwave it.)

Do not waste heat when using the range top. Make sure that pans are flat-bottomed and match the size of the surface unit.

Self-cleaning ovens use less energy when baking or roasting because they are better insulated. To save money, the best time to clean a self-cleaning oven is right after cooking a meal, while the oven is still heated.

When boiling or steaming foods, try using less water than the recipe calls for, and use covered cookware to speed up the process.

Consider all your options. Depending on the size of the meal and the time you have to prepare it, you may save energy by using different appliances, e.g., a slow cooker, fry pan, or microwave.

Microwaves cook foods faster and at a lower wattage, so they often will use less energy than a conventional range.

Whenever you can, run only full loads in your dishwasher, unless it has a partial load setting. You will save electricity and use less hot water.

Use full loads of clothes for both washer and dryer.

Do not overload your dryer. Clothes not only will take longer to dry, but overloading is hard on the dryer.

Clean your washer and dryer filters after every use. Clogged filters mean higher, unnecessary operating costs.

Avoid overdrying clothes. Overdrying wastes energy, damages fabrics, and increases ironing time.

Use the correct setting for the size of the load you are washing. If it is a small or medium load, use that setting.

Your water heater temperature can be set as low as 120°F. If you have an automatic dishwasher, the temperature should be set at 140°F.

Insulate your water heater and hot water pipes to prevent heat loss. You will improve efficiency and reduce hot water costs.

Keep your faucets dripless. Leaky hot water faucets that drip once per second will waste about 2500 gallons per year. Replace faucet washers that drip, and you will save water and energy.

When washing clothes, use cold, cool, or warm water whenever possible.

Lights are important for safety, security, and sight. And while there are energy-saving things to be aware of, keep in mind that lighting makes up only a small portion of your electric bill.

Look for bulbs that deliver the most lumens at the lowest wattage. Light output is measured in *lumens* (shown on light bulb packages). Be sure you get all the output of light you can.

Wherever possible, substitute fluorescent lighting for incandescent lighting, especially in the kitchen, bathroom, and utility areas. You can now adapt most portable lamps, and some ceiling sockets and hanging fixtures to fluorescent lights. Ask for a color-corrected type for a more pleasing atmosphere.

When economical, use a "watt saver" type of incandescent bulb that gives the same amount of light as a standard bulb but uses less energy.

Avoid long-life bulbs because they are the least efficient of the incandescent bulbs. Long-life bulbs cost more to buy and provide less light. They should be used only in areas where bulbs are difficult to replace, such as high ceilings, hallways and stairs.

Use one higher-wattage bulb instead of several lower-wattage bulbs. For example, one 100-W bulb gives the same amount of light as two 60-W bulbs but uses 17 percent less energy. However, be sure you do not exceed the manufacturer's recommended wattage for the fixture.

In areas that do not need to be greatly illuminated, such as hallways or storage areas, use lower-wattage bulbs.

Keep bulbs and fixtures clean and free from dust. Dirt absorbs light and reduces efficiency.

Replace darkened light bulbs before they burn out. As bulbs darken, they consume the same amount of energy but produce less light.

Use light colors on walls, ceilings, and floors to reduce lighting costs. Light-colored rooms reflect more light so they can use lower-wattage bulbs.

Move the refrigerator if it is currently located near the stove, dishwasher, or heat vents. Vacuum the coils every 3 months; dirt buildup makes the machine work harder to keep its contents cool. Check the door gaskets for air leaks. If ice buildup in the freezer is more than ¼ inch thick, defrost.

Scrape but do not prerinse your dishes by hand if you have a dishwasher that automatically prerinses or has a rinse/hold cycle. Machines with these features are designed to dispose of all food particles. Using the "energy saver" option found on many machines can reduce the energy needed to wash a load of dishes and save time and water.

Preheat your oven only when the recipe calls for it, and turn off the oven shortly before the recipe suggests. The heat in the oven can finish the job.

Cook in pots that fit the size of your stove top burners to cut energy waste. Using lids on your pots and pans means you can lower the temperatures and reduce the energy used.

Match the water level and temperature settings on your clothes washer to the size of your load. Do not fill the tub for just a few small items. Follow the manufacturer's directions for other energy-saving hints.

Remember to clean your clothes dryer filters after each use or as necessary.

Lower the temperature setting on your water heater. Many thermostats are preset at the factory at 140°F. Lowering it to 120°F will save you 15 percent of your water heating energy.

Energy Savings, Codes, Standards, and Safety

5

Selecting a method or material to achieve maximum energy savings many times has to be evaluated in light of local and national codes as well as from the perspective of safety. Often it is necessary for the home builder or home owner to seek a reference or expert in a particular field or discipline. There exist numerous agencies, institutes, and organizations from which information and standards can be obtained on safe, energy-efficient building construction. This chapter provides listings for several of these code and standard-setting agencies, their addresses, telephone numbers, and a brief description of what each one provides.

Besides all the code and standard-setting agencies that can provide references for home owners and home builders, there are many other guidance vehicles. The U.S. Department of Energy (DOE), the U.S. Environmental Protection Agency (EPA), and the Federal Trade Commission (FTC) have gotten very involved in energy savings and conserving our earth's resources. Not only can the home owner save valuable energy dollars by following their guidance, but also the ultimate saving is in not having to continue producing energy at a rate that eventually will use all the earth's resources and pollute the atmosphere. Very strict laws, guidelines, and regulations have been set in place over the past few years by the governing bodies that will force manufacturers and builders to "get in line."

EPACT

One set of regulations is EPACT. EPACT, the *Energy Policy ACT* of 1992, was introduced and finally enacted so as to minimize the dependence of the United States on foreign oil for security reasons, reduce the environmental impacts of energy production, and maintain the technological competitiveness of the United States. EPACT made an impact. While there have been numerous energy-related acts and regulatory guidelines over the past 20 years, EPACT is the most far-reaching and the most quoted (much more so than the Clean Air Act). EPACT will be enforced, and thus users and manufacturers are taking action. Electric motors have to meet certain new and higher-efficiency ratings. All state and government facilities, even those located abroad, were to begin programs that would save that facility energy in the short and long term. Electronic ballasts and variable-frequency drives were to be used wherever possible. This energy act also allowed, for the first time, wholesale generators of electricity to apply to the FERC (Federal Electrical Regulatory Commission) to order the transmission facilities of other companies to grant them access so that they could send their power to other wholesale generators (Section 721). This did not apply to wholesale customers accessing transmission facilities to send electricity to retail customers, since this is typically within the jurisdiction of state regulators. However, deregulation of the electrical power companies is now a reality at the retail level.

EPACT covers energy efficiency, electric vehicles, renewable energy, and alternative fuels. The process of its implementation, culminating with the issuance of FERC Orders 888 and 889, has brought on a rapid rearranging of corporate and regulatory institutions within the industry. Congress wanted to create an electric power market that would place greater reliance on competition, discipline prices, and bring out the greatest overall economic efficiencies that utilities can achieve.

Energy Star Program and EnergyGuide Labels

Both these programs, produced in cooperation with the DOE, are major steps in helping consumers make the energy-efficient, correct choice. The EnergyGuide Program has been in existence for several years and is primarily a consumer guide. With the federal government backing the program, manufacturers of appliances have had to adhere to more exact energy standards. The Energy Star Program is taking the energy savings and energy-efficiency standards to another level. It provides for labeling and manufacturing standards that ensure that residences and products are going to save energy in the future. One major outcome of both programs is that if less energy is needed by consumers, then less will have to be produced by the power plants—thus helping to save the environment.

Frequently, consumers purchase appliances and electrical products based on sales and marketing tactics. Today, by federal law, all major residence appliances must meet federal minimum energy-efficiency standards set by the DOE. One promising by-product of this program is that many appliances beat the standard and use even less energy. These energy programs provide the consumer with very real data (from which to make an educated decision), and manufacturers must use standard test procedures developed by the DOE to prove the energy use and efficiency of their products.

Many manufacturers will have these tests performed by independent laboratories. The test results are reported on the EnergyGuide.

The yellow EnergyGuide label appears on practically all electrical appliances. Hot water tanks, washing machines, dryers, furnaces, dishwashers, and many more products will carry the label. Some appliances such as microwaves and electric ranges do not have to carry a label because their energy differences are very minor between models. Most of the actual energy savings and the differences are in the internal makeup of the product. Gasketing and sealing, pumps, electric motors, compressors, and valving all are areas within a product where better efficiencies can be attained. By using higher-efficiency motors or better gasketing, a product's overall efficiency goes up.

An example label from an electric hot water tank (tank A) is shown in Fig. 5.1. It compares the energy use of competing models, estimates their differences in energy costs, and allows the consumer to consider both purchase price and estimated energy use when deciding which brand and model to buy. Another label from hot water tank B is shown in Fig. 5.2. Here's how to interpret the label: There is an EnergyGuide heading, with an arrow pointing to either the highlighted estimated yearly energy cost or the energy-efficiency rating of the appliance. Furnaces will carry only a general label and will have separate fact sheets available from the furnace dealer. There will be a range of estimated yearly energy costs or energy-efficiency ratings of competing models within the same product category. A statement that energy cost varies with the energy rate and individual use patterns also appears on the label. There is a table showing the energy cost of the appliance based on a variety of utility rates and how the appliance is used by the consumer. Remember, this label is only a guide; it is not in any way a guarantee.

Model A Electric Hot Water Tank purchased for $275
Yearly Operating Cost: $589

Model B Electric Hot Water Tank purchased for $395
Yearly Operating Cost: $475

First, subtract the purchase price of the less expensive model (A) from the price of the more expensive model (B):

$$\$395 - \$275 = \$120$$

Next, subtract the lower yearly operating cost ($475) from the higher yearly operating cost ($589):

$$\$589 - \$475 = \$114$$

Water heater Electric   Model XYZ-123
First hour rating 81

ENERYGUIDE

Estimates on the scale are based   Only models with first
on the National average electricity   hour ratings of **75** to
rate of $.0804 per Kilowatt-hour   **86** gallons are used on
                                          this scale

Model with lowest   Model with highest

energy cost: **$429**   **$589**   energy cost: **$614**

Yearly Cost

Cost per	0.04	$292
Kilowatt-	0.06	$438
Hour	0.08	$584
	0.10	$730
	0.12	$876
	0.14	$1022

Ask your saleperson or local utility for the energy rate in your area.

IMPORTANT:Removal of this label before consumer purchase is a
violation of Federal Law (42 IU.S.C. 6302)

■ **5.1** *Sample EnergyGuide label for electric hot water tank A.*

Last, divide the difference in purchase price (step 1) by the difference in yearly operating cost (step 2) to come up with the number of years it would take model A to make up the difference in the initial purchase price:

$$\frac{\$120}{\$114} = 1.1 \text{ years, or approximately 13 months}$$

This is a good payback and probably justifies the premium and the purchase. Also, the higher-efficiency water tank will pay for itself even quicker if the cost per kilowatt hour is higher than the rate used for the sample.

Water heater <u>Electric</u>   Model <u>XYZ-123</u>
First hour rating <u>81</u>

ENERYGUIDE

Estimates on the scale are based
on the National average electricity
rate of $.0804 per Kilowatt-hour

Only models with first
hour ratings of **75** to
86 gallons are used on
this scale

Model with lowest

energy cost: **$429**

$475

Model with highest

energy cost: **$614**

Yearly Cost

Cost per		
	0.04	$235
Kilowatt-	0.06	$353
Hour	0.08	$471
	0.10	$588
	0.12	$706
	0.14	$824

Ask your saleperson or local utility for the energy rate in your area.

IMPORTANT:Removal of this label before consumer purchase is a
violation of Federal Law (42 IU.S.C. 6302)

■ **5.2** *Sample EnergyGuide label for electric hot water tank B.*

Here are some other points to ponder when shopping and comparing
energy-consuming products:

1. Consider other factors such as the manufacturer, selection available,
   price, reputation or reliability, and warranties.

2. Look for the best combination of performance, efficiency, convenience,
   and size.

3. If the unit uses water, how much water does it use?

4. Are any other rebates or offers associated with one particular model over another?

5. Use the EnergyGuide labels.

6. Learn everything that you possibly can about the product. Ask many questions.

The Energy Star Program is much more far reaching than the EnergyGuide Program. It addresses not only various appliances but also the entire house. Besides the label, which is designed to help consumers recognize energy-efficient products, Energy Star offers residence calculation software for energy savings, financing, and mortgages, along with strict standards for the entire residence and its energy-savings ability. The EPA created the Energy Star label as the symbol for energy efficiency because of the direct link between energy consumption and air pollution problems. The more energy consumed, the more coal, oil, and gas have to be burned to keep up with the demand. This adds to the problems of global warming, acid rain, and smog and contributes to the destruction of the ozone layer. By properly sizing and installing high-efficiency heating and cooling equipment, we can lower utility bills and improve the comfort and air quality in our residences, all the while reducing air pollution.

The Energy Star Program has a labeling element similar to the EnergyGuide Program. Again, the Energy Star Program is more detailed and further addresses air conditoners, thermostats, and furnaces. Furnaces that are 90 percent efficient, i.e., having an annual fuel utilization efficiency of 90, or greater qualify for an Energy Star label. Manufacturers of high-efficiency furnaces voluntarily use the Energy Star label on qualifying equipment and related product literature. Likewise, central air conditioners that have a seasonal energy efficiency ratio (SEER) of 12 or greater qualify for the Energy Star label. Seasonal energy efficiency ratio is a measure of cooling efficiency. The larger the number, the greater is the efficiency. Manufacturers of high-efficiency central air conditioners can use the Energy Star label voluntarily on qualifying equipment and related product literature.

Thermostats that are labeled by Energy Star must be programmable thermostats. They must have separate weekday and weekend programs, each with up to four customized temperature settings—two for occupied and in-use periods and two for energy-saving periods when the house is unoccupied or everyone is asleep. Some have an advanced recovery feature that can be programmed to reach a desired temperature at a specific time in a way that minimizes system "on" time and auxiliary heat use. They also must have the ability to maintain room temperature within 2°F of the desired temperature. A hold feature must be included that allows users to temporarily override automatic settings without deleting programs. (For example, programming can be adjusted to maximize savings during a vacation or extended absence.)

Energy Star also provides for the entire residence to have an overall efficiency rating. The *Model Energy Code* (MEC) is a national standards-setting entity that has set energy-efficiency guidelines for residences in various states and is discussed later in this chapter. The Energy Star program maintains that a house must be built at least 30 percent more efficient than the current national MEC. The 30 percent reduction actually equates to a Home Energy Rating System (HERS) rating of at least 86 on the HERS point scale. A HERS rating is an evaluation of the energy efficiency of a residence compared with a reference house (same size and shape as the rated residence) that meets the requirements of the national MEC. The most cost-effective improvements tend to be efficient hot water heating, tightly sealed and insulated ducts, air-leakage control, high-performance windows, and high-efficiency heating and cooling equipment. Energy Star levels can be reached without increasing costs. However, the reality is that Energy Star home features can increase the value of a residence and increase the purchasing power of buyers. Such features can help you differentiate one residence from another.

NAECA

In 1987, to promote efficiency and conservation, Congress passed the National Appliance Energy Conservation Act (NAECA), which mandates that manufacturers make use of readily available technologies in building appliances 10 to 30 percent more efficient than previous models. Since these new refrigerators, air conditioners, water heaters, and lighting systems use less energy to perform as well as or better than their predecessors, the American consumer benefits from lower monthly utility bills while sacrificing nothing in terms of reliability or comfort. Analysts estimate that NAECA and subsequent legislation will save consumers more than $200 billion—or about $2000 per residence.

The standards set by NAECA and its subsequent revisions are benchmarks for how much energy a particular appliance needs to use. By law, manufacturers must provide detailed yellow EnergyGuide labels on all new appliances for sale. Drawing on data from standardized government tests, these labels provide estimates of an appliance's yearly operating costs and efficiency levels in comparison with competing models. A consumer who compares these energy-efficiency rating (EER) statistics should have all the information needed to calculate the long-term energy costs—not just the short-term purchase expense—of items under consideration. Consumers who buy efficient appliances benefit because they save dollars on their utility bills. Refrigerators made to meet the latest DOE efficiency standards, which were issued in April of 1997 and will take effect in the year 2001, will cut consumers' energy costs by 30 percent compared with those manufactured to the previous standards, set in 1993. Table 5.1 shows appliance expected life.

■ **Table 5.1 Life Expectancies
of Various New Appliances**

Type of Appliance	Expected Lifetime, Years
Electric ranges	18–20
Refrigerators	15
Freezers	18–20
Dishwashers	11
Clothes washers	11
Water heaters	15–18
Room air conditioners	12–15
Furnaces	20

NOTE: The average life expectancy will vary depending on the actual maintenance performed over the lifetime of the appliance.

Demand-Side Management

Demand-side management (DSM) can be looked on either traditionally, as a tool to be used to change the demand for energy, or more generally, as a tool for society to better use and distribute scarce resources. When implemented by energy utilities, this change might be in the amount of energy used or the pattern of its use. For example, an electric utility may increase the power supply and provide customers with the kilowatthours they ask for or provide better equipment that uses fewer kilowatthours while providing more service. Or the utility company may seek to change the pattern of energy use by applying straightforward and operationally simple time-of-use pricing to shift the load. In either case, the objective is to choose the best option to provide energy services at the lowest cost.

Demand-side management as a more general tool opens up a broader range of long-term activities that are triggered by society's concern for the environment and desire for sustainable development. This option can be more complex, and it is often supported by other goals and activities, e.g., market transformation and opportunities for the emergence of new types of industry. In both cases, the main thrust and reason for DSM are the need to increase energy efficiency and receive better value for the capital invested in the energy system.

Codes

Many standards exist, as well as many regulations for proper and safe building construction. There are standards, rules, or models to which other related things are compared, and then there are the codes, the set of laws for a particular method or process. The codes will take a different kind of importance over the standards, but frequently the standards can be used in most code situations. Codes demand compliance, or else fines and penalties can be levied, whereas not following a standard may not necessarily result in increased cost. For instance, the standard is to use solder (with or without lead) to connect a copper pipe to another copper pipe. In a particular locality, the *only* acceptable solder is lead-free; therefore, there is no choice. Thus local and state building codes must be checked and followed. The Chamber of Commerce in a locality can provide these as well as many of the utilities in a certain region, since they trade and exist in that local area.

Once a particular organization has lobbied the federal government and one of its legislative branches to get a regulation to the level of code, then that code is mandatory throughout the country. One such code is the *Model Energy Code*. The MEC addresses several aspects of a residence's energy efficiency. Compliant materials, insulation, windows, and doors are just a few of the areas required to meet the code. The DOE provides support for the MEC by developing compliance materials (called *Model Energy Codecheck*) that simplify its use and by promoting training. The Model Energy Codecheck Software (IBM-compatible) does all the necessary code-compliance calculations. The Model Energy Codecheck Prescriptive Packages show insulation and window requirements at a glance. The *Model Energy Codecheck Manual* includes forms, checklists, and worksheets for documenting compliance. The Model Energy Codecheck materials can be downloaded at no cost.

States that have adopted the MEC are shown in Table 5.2. As can be seen, there are still quite a few states that could adopt the MEC. The MEC also can be adopted by counties and cities in states that do not have a statewide energy code. In those areas where the MEC is in effect, a building permit cannot be obtained until the MEC application is approved. Residences financed by the FHA and the Veterans Administration must comply with the 1992 *Model Energy Code*. The MEC was enacted to promote energy requirements for new residences. The U.S. Department of Housing and Urban Development (HUD) loan guarantee program requires compliance with the MEC. The MEC applies to new residential buildings of three stories or less in height. The MEC is designed to provide for energy-saving designs in insulation use and the amounts to make the residence energy

■ **Table 5.2 States that Have Adopted the *Model Energy Code* (through 1997)**

State	Year Adopted
Tennessee	1992
Arkansas	1992
Iowa	1992
Indiana	1992
New Mexico	1992
Ohio	1993
Virginia	1993
Delaware	1993
Montana	1993
North Dakota	1993
Utah	1993
Kansas	1993
Georgia	1995
Massachusetts	1997
Maryland	1997

efficient. Single-family homes account for about 15 percent of the total energy used in the United States. Better efficiencies in these homes hopefully will lower this percentage. Also, greater energy savings in homes lead to increased value for builders in marketing their homes.

Two other codes that play an important role in energy efficiency within homes are the *Uniform Plumbing Code* (UPC) and the *Uniform Mechanical Code* (UMC). The UPC and UMC contain several sections that are useful and necessary to obtain energy savings. The most important energy-related portions of the UMC are General Requirements for Heating, Ventilating and Cooling, Ventilation, the Air Supply and Exhaust Systems, Product Conveying Systems, Special Fuel Burning Equipment, Duct Sizing Tables, Duct Systems, and the following sections:

Section 603: Installation

Section 604: Insulation

Section 605: Dampers

Section 607: Plenums

Likewise, the UPC addresses the material types and accepted methods for all the in-house piping and plumbing. Most heating and cooling systems require close adherence to sections of the UPC, and thus the two codes often overlap.

Safety and Energy Savings

Making the residence safe while reducing energy consumption should be a routine and simple task; however, in one way or another, some mishaps may occur. Safety relative to energy savings might mean sealing the house so airtight that there is a health and breathing hazard as a result. Or if too many wires are run at the wrong gauge, or thickness, a fire hazard could present itself. Even a fully automatic home automation system could fail if the conditions are right.

Testing agencies such as Underwriter's Laboratories (UL), the Electronics Testing Laboratories (ETL), and the Canadian Standards Association (CSA) all perform rigorous tests on newly introduced equipment. While they test under many circumstances and under severe loads and are looking for all types of problems, the most important check is to see that no sparking, smoking, or fires can occur. Other organizations such as the National Fire Protection Agency (NFPA) and the *National Electric Code* (NEC) also are continually adapting to the changes in technology so as to ensure safety. As a matter of fact, all standards organizations hold safety above all else, when setting or maintaining guidelines. For the most part, energy efficiency and energy savings can be attained quite adequately without sacrificing safety. In some rare and extreme instances, accidents can happen, but with enough checks and balances in the "energy system," these are few.

Standards Organizations

Air-Conditioning and Refrigeration Institute (ARI)
4301 North Fairfax Drive
Suite 425
Arlington, VA 22203
(703) 524-8800

The Air-Conditioning and Refrigeration Institute (ARI) is the national trade association representing manufacturers of more than 90 percent of U.S.-produced central air-conditioning and commercial refrigeration equipment. The ARI's history dates back to 1903, when it started as the Ice Machine Builders of the United States. The ARI was formed in 1953 through a merger of two related trade associations. Since that time, several other related trade associations have merged into the ARI, making it the strong association that it is today. Over the past 45 years, the ARI has emerged as the major voice for the air-conditioning and refrigeration industry.

Much like many of the other organizations listed here, the ARI develops and publishes technical standards for industry products. The ARI standards establish rating criteria and procedures for measuring and certifying product performance in the industry and its products. The products are rated on a uniform basis so that buyers and users can properly make selections for specific applications. Standards are developed by individual ARI product sections and other interested parties who wish to participate and then are approved by the ARI's General Standards Committee. The more than 60 standards now published are mainly performance-rating standards, although some are application or terminology standards. Many of the ARI standards are accepted as American National Standards. The ARI actively participates in developing international standards and has established a policy of adopting international standards for use in the United States when practical and feasible. The ARI is cooperating with the Canadian Standards Association (CSA) and other groups to establish joint ARI/CSA rating standards and common U.S./Canadian safety standards.

The ARI verifies manufacturers' ratings of products in continuous and extensive laboratory testing. Each product section, with the support of the ARI engineering staff, may develop certification programs for its eligible products. Participation in the programs is voluntary and is open to non-members of the ARI on an equal basis. The ARI annually selects a significant number of each manufacturer's production models to be tested by an independent laboratory under contract to the ARI. Randomly selected units are tested using procedures stipulated in the corresponding ARI standards to verify that they meet the manufacturer's certified published performance ratings. A test failure requires rerating or cessation of production of the failed product.

Membership in the ARI is available to any corporation or firm in North America that manufactures and sells in substantial volume in the open market any product within the product scope of one or more ARI product sections. International membership is available to manufacturers abroad that sell any product within the scope of one or more ARI product sections in North America. Associate membership is available to manufacturers who sell in substantial volume any product that is used in the production or operation of products within the product scope of one or more ARI product sections and which is not otherwise eligible for membership.

American Society of Heating, Refrigeration, and Air-conditioning Engineers (ASHRAE)
   1791 Tullie Circle, N.E.
   Atlanta, GA 30329
   (404) 636-8400

ASHRAE is probably the most recognized standards authority in the United States regarding heating, ventilating, refrigeration, and air conditioning.

Extensive study and work have been done by this organization for many years. Thus it is important to include this group whenever discussing energy efficiency and energy systems. The two standards that are most relevant here are ASHRAE 90.1-1989, Energy Efficiency for Commercial and High Rise Structures, and ASHRAE 90.2-1989, Energy Efficiency for Low Rise Construction. Both these standards contain numerous guidelines for energy-efficient building construction. Copies of these standards can be purchased from ASHRAE at the address given.

At present, there is a proposed revision to ASHRAE Standard 90.1. It seeks to encompass a variety and series of energy-efficiency requirements. Since buildings and their systems differ considerably in age, size, location, and climate, the Building Owners and Manufacturers Association (BOMA) International opposes a "one size fits all" approach. Systems installed in existing buildings should not be expected to meet the same energy-efficiency levels as can be designed into new buildings. The current edition of ASHRAE Standard 90.1, crafted in 1989, is the most widely referenced set of guidelines on this subject. It was adopted by the federal government in the Energy Policy Act of 1990 and subsequently included in most state building codes. The revised standard would affect a wide scope of construction, including new buildings, new portions of facilities, and new systems or equipment installed in existing structures. These energy-efficiency requirements would apply to the building envelope as well as to a variety of systems, including heating, ventilation, and air conditioning (HVAC), service water heating, electric power distribution and metering, and lighting. The overall intent of this standard is to replace inefficient equipment with higher-efficiency equipment, sometimes regardless of cost, which has helped to make the revision somewhat controversial.

American Institute of Motion Engineers (AIME)
Kohrman Hall
Western Michigan University
Kalamazoo, MI 49008
(616) 387-6533

This organization is made up of members from many disciplines, including electrical, mechanical, and computer professionals. All members have one focal point—motion and motion control. Any equipment that uses a prime mover, be it an electric motor, hydraulic system, pneumatic system, or other, has a moving element to it. The understanding and control of this motion are the goals of AIME. Energy efficiency is studied in a twofold manner by this group: (1) the actual efficiency of the prime mover (hence premium-efficiency motors) and (2) incorporation of these motors and prime movers into fan or pump systems used to save energy overall.

American National Standards Institute (ANSI)
11 West 42nd Street
New York, NY 10036
(212) 642-4900

The American National Standards Institute (ANSI) has been in existence for many years. It is the body by which many standards, over the years, have gained credibility. As the "secretariat to many standards," ANSI's involvement over a wide range of disciplines is often sought.

Building Owners and Manufacturers Association (BOMA)
1201 New York Avenue, NW
Suite 300
Washington, DC 20005
(202) 408-2662

The Building Owners and Manufacturers Association (BOMA) is mainly an organization for commercial building. However, because so many residential practices and standards arise out of the industrial and commercial marketplaces, an organization such as BOMA is valuable to the home builder and home owner. The primary mission of BOMA is to actively and responsibly represent and promote the interests of the commercial real estate industry through effective leadership and advocacy; through the collection, analysis, and dissemination of information; and through professional development.

BOMA International was founded in 1907 as the National Association of Building Owners and Managers. The association assumed its present name in 1968 as it broadened its reach to include Canada and participants from around the globe. Today, BOMA International represents 84 U.S., 10 Canadian, and 5 overseas associations in Australia, Japan, Korea, the Philippines, and South Africa. BOMA's North American membership represents a combined total of more than 6 billion square feet of office space. BOMA provides a network forum for industry professionals to discuss mutual problems, exchange ideas, and share experience and knowledge. BOMA offers comprehensive industry research and office building performance data and publications, products, and services to assist the owners and managers of commercial office space to become more expert, professional, and accomplished in creating better tenant environments and more stable investments for world capital. BOMA has been the standard-setter for the office building industry and is often a guide for residential construction. In 1916, the first Standard Chart of Accounts was issued. Since then, it has been revised four times: in 1939, 1976, 1988, and 1998. In 1920, BOMA began disseminating extensive industry research. The breakthrough publication was the *Experience Exchange Report* (EER), reporting on office building incomes, expenses, and overall performance. Publication of the EER marked the first time building owners and managers had been willing to share information on the operation of their buildings. It has been published every year since. In 1997, over 4000 North American office buildings representing more than 794 million square feet reported from more than 126 cities.

In recent years, BOMA has expanded its marketplace research capabilities with tenant, compensation, custodial, and other surveys and access to "how to" guidebooks and information-packed publications. BOMA's government affairs and codes representation addresses the needs and interests of office building owners, investors, developers, and managers to national, state, provincial, and local legislators. Since its inception, BOMA has been politically active representing the interests of the industry. In 1929, BOMA served on the National Conference on Business called by President Hoover to assess business conditions. In 1933, BOMA won a wage and price provision in the National Recovery Act to protect building owners from disastrous cost increases. During World War II, BOMA served on the National Coal Board to allocate fuel supplies.

BOMA found it important to be closer to the federal government during the 1970s because of the crunch of legislative and regulatory activity affecting office buildings. In moving to the nation's capital, BOMA immediately took on the Capital Cost Recovery Act, building energy performance standards, energy tax credits, and fire safety. BOMA offers continuing education programs in a variety of current industry topics, including its popular "Boot Camp," an intensive 3-day course for property managers, and "Top Gun," an advanced skill-building program for "Boot Camp" graduates. BOMA's awards program recognizes excellence in building management, operational efficiency, tenant retention, emergency planning, and community impact. BOMA pays homage to the buildings it represents, to the individuals it serves, and to its federated associations through the Office Building of the Year "TOBY" Award. This comprehensive program recognizes quality in office buildings and awards excellence in office building management. A building's operations are thoroughly evaluated in terms of tenant relations programs, community involvement, emergency evacuation procedures, and continuing education programs for personnel.

Council of American Building Officials (CABO)
  5203 Leesburg Pike
  Falls Church, VA 22041
  (703) 931-4533

The Council of American Building Officials (CABO) publishes the *Model Energy Code*, and its Code Changes Committee updates it annually. New editions of the MEC are published at approximately 3-year intervals.

Canadian Standards Association
1200 - 45 O'Connor
Ottawa, Ontario K1P6N7
(613) 328-3222

The Canadian Standards Association (CSA) is probably the most recognized standards organization in Canada. It has as much recognition as UL and ANSI in the United States, and its standards must be adhered to by U.S. and foreign manufacturers in order to have non-Canadian products enter into Canada.

Energy Efficient Lighting Association (EELA)
P.O. Box 727
Princeton Junction, NJ 08550
Phone: 609-799-4900   Fax: 609-799-7032

The Energy Efficient Lighting Association (EELA) is an organization that lists and promotes the use of all energy-efficient lighting technologies. The entire lighting marketplace of manufacturers, distributors, contractors, designers, architects, energy service companies, building owners, and facility managers benefits from EELA's program. Through its awareness and educational campaign called "Enlightening America . . . It's Good Business," EELA promotes greater understanding, networking, and cooperation among all segments of the lighting industry to help get lighting upgrade projects installed sooner.

Energy Efficiency & Renewable Energy Clearinghouse (EREC)
U.S. Department of Energy, EREC
P.O. Box 3048
Merrifield, VA 22116
1-800-DOE-EREC
TDD: 1-800-273-2957
http://www.eren.doe.gov

This organization is a branch of the U.S. Department of Energy. This branch deals strictly with energy-efficient products and standards as well as renewable energy systems. Documentation and publications concerning energy efficiency and products, many of which are free, are available at the above-mentioned address or Web site.

U.S. Department of Energy (DOE) Headquarters
Forrestal Building
1000 Independence Avenue, S.W.
Washington, DC 20585

The U.S. Department of Energy is a federal agency. The DOE basically gets involved with virtually any issue relating to energy. It governs, controls, and sets legal guidelines for all energy sources: petroleum, natural gas, fossil fuels, and nuclear energy. The DOE also addresses renewable and "green" power sources such as solar, biomass, hydroelectric, and so on. A good deal of research is funded through this agency, and generally, results and statistics are made available to the public.

Electrical Apparatus Service Association, Inc. (EASA)
1331 Baur Boulevard
St. Louis, MO 63132

The Electrical Apparatus Service Association (EASA) is comprised of manufacturers, motor repair and service groups, and other professionals involved with the repair and service of electrical equipment. Electrical motors, starters, and other motor controls have to meet exacting guidelines to satisfy the new energy standards such as EPACT and others. This group maintains this sector of the discipline.

Electronic Industries Association (EIA)
2001 Pennsylvania Avenue
Washington, DC 20006-1813
(202) 457-4919

This group is made up of manufacturers and professionals from the electronic products disciplines. With so much equipment being electronically and microprocessor based, this organization is in place to set guidelines and monitor and assist in the field.

Electronics Testing Laboratories, Inc. (ETL)
Industrial Park, P.O. Box 2040
Cortland, NY 13045
(607) 753-6711

This private agency is able to perform laboratory testing and certification for most electrical and electronic products. Electronics Testing Laboratories (ETL) is an independent third-party testing organization whose label is recognized around the country. Many energy-efficient products are tested and certified by ETL.

Federal Trade Commission (FTC)
Public Reference
Washington, DC 20580
(202) 326-2222 TDD: (202) 326-2502
http://www.ftc.gov

The Federal Trade Commission (FTC) is an independent agency that seeks to protect the public against unfair, deceptive, and fraudulent advertising and marketing practices. Many publications are made available to the public at the above-mentioned address and Web site.

Home Energy Rating Systems (HERS) Council
1511 K Street NW, Suite 600
Washington, DC 20005
(202) 638-3700   Fax: (202) 393-5043

The Home Energy Rating Systems (HERS) Council is a nonprofit association whose mission is to increase residential energy efficiency and affordability nationwide by advancing uniform home energy ratings, energy-efficiency financing, education, and research through collaborative efforts. The council maintains that once the linkage between home energy ratings and energy-efficiency financing is established, energy-efficient homes will become more available and affordable.

Institute of Electrical & Electronic Engineers (IEEE)
445 Hoes Lane
Piscataway, NJ 08854
(908) 562-3803

The Institute of Electrical and Electronic Engineers (IEEE) is a society of professionals and manufacturers that has been in existence for many years. IEEE addresses and sets guidelines for hundreds of topics relating to electricity and electrical phenomena. Extensive research and its publication are done by the IEEE. Many articles, papers, and publications are available for a fee.

National Association of Home Builders (NAHB)
1201 15th Street NW
Washington, DC 20005
(800) 368-5242

The National Association of Home Builders (NAHB) is the most influential association representing the housing industry. Working in partnership with nearly a thousand state and local builders' associations nationwide, NAHB provides a wide range of services to its members and is a powerful advocate for housing and the home building industry.

Home Builders Institute (HBI)
1090 Vermont Avenue, NW, Suite 600
Washington, DC 20005
(202) 371-0600
Fax: (202) 898-7777

The Home Builders Institute (HBI) is the nation's leading source for education and training programs serving the home building industry. For more than 25 years, HBI has trained skilled workers in residential construction. HBI helps builders enhance their professionalism through continuing education and certificate programs. In 1983, HBI was officially named the educational arm of the National Association of Home Builders.

> National Council of the Housing Industry (NCHI)
> 1201 15th Street, NW
> Washington, DC 20005
> 800-368-5242, ext. 520
> 202-822-0520

The National Council of the Housing Industry (NCHI) is a special standing committee of NAHB that provides the nation's leading suppliers of building products and services the opportunity to participate in all the NAHB's activities, affect housing issues, and gain direct access to NAHB's builder members. Founded in 1964, it consists of companies that supply products and services and related industry trade associations representing the core leadership of the building industry. These members are trusted and recognized leaders in the construction field who have earned reputations for reliable service and innovative products for the building industry. As members of NCHI, they subscribe to the highest product and service standards, setting them apart from other manufacturers and service organizations. NCHI members provide a unique resource for assisting in problem solving on a variety of issues related to providing better housing for Americans. Members work closely with NAHB's senior officers—the key decision makers for national housing policy—to ensure that the interests of suppliers are properly conveyed and considered during key policy development activities.

> National Electrical Manufacturers Association (NEMA)
> 2101 L Street, NW
> Washington, DC 20037
> (202) 457-8400

The National Electrical Manufacturers Association (NEMA) is used in technical discussions quite often. Made up of members from manufacturers, engineering, and academia, NEMA sets standards for practically everything electrical. Motors, drives, starters, and enclosures are just a few of these products. NEMA has done extensive research and testing of electrical products and can provide backup documentation on request.

National Fire Protection Association (NFPA)
1 Batterymarch Park
P.O. Box 9101
Quincy, MA 02269-9101
1-888 NEC Code (1-888-632-2633)

The National Fire Protection Association (NFPA) is the authority on fire and electrical safety for industry, commercial, and residential applications. NFPA publishes many books and brochures on fire protection, as well as the *National Electrical Code*. The NFPA has been in existence well over 100 years, and its publications and other works are accepted by the American Insurance Association and the Fire Underwriters. The *National Electrical Code* is revised and reprinted every 3 years and is supposed to be purely advisory, as far as the NFPA and ANSI are concerned, but is used extensively in regulatory systems and in legal matters whenever life and property protection are the issues. The *National Electrical Code* and *NEC* are registered trademarks of the NFPA.

National Institute of Standards & Technology (NIST)
Building 221/A323
Gaithersburg, MD 20899
(301) 975-2208

The National Institute of Standards and Technology (NIST) was established by Congress to assist industry in the development of technology. An agency of the U.S. Department of Commerce's Technology Administration, NIST's primary mission is to promote U.S. economic growth by working with industry to develop and apply technology, measurements, and standards.

Society of Manufacturing Engineers (SME)
One SME Drive
P.O. Box 930
Dearborn, MI 48121
(313) 271-1500

The Society of Manufacturing Engineers (SME) is an international professional society dedicated to serving its members and the manufacturing community through the advancement of professionalism, knowledge, and learning. SME was founded in 1932 and has several thousand members in over 70 countries.

Underwriters Laboratories (UL)
333 Pfingsten Road
Northbrook, IL 60062
(708) 272-8800

This is a third-party testing agency that has been around for decades. The UL label is recognized around the country, and even around the world, as a symbol of product integrity and safety. Underwriter's Laboratories tests all types of products extensively and also sets the testing standards for which the tests are governed. Many residential products carry the UL label on them.

Measuring Energy Savings

6

The builder, contractor, and home owner all must have the means to determine, on paper, if and where energy savings can be obtained. This means that a method for designing and building around energy savings must be addressed and then a system should be implemented to substantiate these predicted energy savings later. This is easily stated but difficult to incorporate! Maybe this is why energy savings is such a challenge. Where are the heat losses, the wasted energy, and equipment within the home that are energy-hogs? Sometimes it is as simple as replacing a 40-year-old, inefficient furnace, and other times we must methodically find energy savings with every insignificant component in the house. It all adds up.

There is no single answer, even though there is one standard question: Why can't I just have the most energy-efficient house available built? You can, but each home owner's needs, wants, and situation must be evaluated. Not every house is built the same, and not every house contains the same appliances. Therefore, it is necessary for everyone to understand energy efficiency and be able to quantitatively determine what is a good energy system and what constitutes a bad energy system. This chapter will help you to calculate, predict, and measure energy savings.

If, Where, and How Much? The First Energy Savings Questions

We cannot realize any energy savings until we know if, where, and how much energy we are losing or will lose. Actual or predicted heat losses must be calculated first. Where do we start? What components within the proposed home or retrofit need attention? What components need to be replaced? If we change this, will we realize any energy savings? What is the projected heat loss of the structure? Many questions, and where do we begin? Three important beginning factors are

1. Is this a new, built from the ground up residence?
2. Is the residence existing and going to be retrofit?
3. Will a home automation system be incorporated?

Obviously, starting from scratch with a new site and a new building is the ideal situation. The home builder can provide guidance and use tried and sound practices for energy savings. The home owner can make requests, and the builder can work an energy-efficient design around those requests. This is the time, during design and on paper, that all thoughts and changes can be made without sacrificing energy savings. This is not always the case with a retrofit of an existing residence. We have to work with what we've got—the existing house. This is a harder proposition because walls and ceilings are already in a finished state. This means that methods and techniques for insulating and energyproofing the home are a bit harder. There are blow-in insulation systems that minimize the disturbance to the existing walls, and the attic and basement spaces usually can be accessed for the placement of insulation.

As for home automation systems, a new home is the best situation for installing the wiring and additional components for such a system. It must be determined to what extent automatic control is required, where controls and human interface panels will be located, and how the system will be used for energy savings. Involve a home automation specialist at this time, or have the home builder address these issues. The installation process is critical to the success and use of a home automation system. Existing homes have to be approached in a different manner. Existing homes have to be analyzed by a home automation technician to see how and if a system and its wiring can be installed easily. If not, then a wireless system may have to be considered. A full examination of home automation and wireless systems is presented in Chap. 10.

Where is thermal energy lost in a typical residence? The answer is in many places. Energy-efficient design starts with the walls and roofs. For fuel savings, walls and roofs must be resistant to the rapid transmission of heat. Slow passage of heat also results in warmer, more comfortable inside temperatures. Insulation is probably the best material to incorporate into any wall or roof, and this is evidenced later on. Vapor barriers also need to be incorporated because they help to control the moisture content of the residence and thus assist in temperature control of the interior. The nature of heat flow also must be factored. A reduction in the rate of heat loss can be achieved by the use of insulating materials having slow conduction rates, the introduction of one or more air spaces, and the use of reflective linings to reduce radiant transfer of heat within those air spaces. Greater thicknesses of solid materials help, but this results in less control of internal convection and the inside/outside surface conductance losses.

Whenever a wall or roof is incorporated to enclose a room under conditions resulting in heat loss from that space, the air movement inside and outside the room acts to increase the rate of heat loss from the room. The gentle stirring of the normally still air within the room and the more active motion of the wind outside the room together work against energy savings. These convection currents within the room will cause the heated air molecules to collide with the cooler surfaces. This exaggerates the isothermal planes phenomenon described earlier in this book. The net result is a quicker loss of heat. A greater temperature differential and a higher velocity of the outside winds also act to aggravate the heat loss.

Glass and windows are another major source of heat loss. Glass transmits heat more rapidly than practically any other material except maybe a sheet of metal. Unfortunately, there is little that can be done about this. The effective solutions for windows vary, but the most commonly employed method is that of multiple panes of glass separated by air spaces. Double glazing is better than single, and triple glazing is better yet. However, the cost for the better window also escalates accordingly. The orientation and placement of glass can promote the admittance of solar energy during sunny hours of the day, but heat loss calculations usually are done assuming the coldest periods, typically during the dark hours. Today, triple glazing and glass substitutes are able to deliver some fairly high R values.

Other locations that need to be examined are the attics, crawl spaces, and basements. Unheated crawl spaces are often ventilated with outdoor air during the winter. Typically, the floor above the crawl space is well insulated and airtight. The crawl space is usually treated as an outside element, and the heat loss is calculated from the room to the crawl space, with the outdoor temperature assigned to the crawl space. If the crawl space has vents that can be closed and the space heated, then there is no heat loss through the room floor, and the heat loss from the crawl space is treated as though the crawl space were a basement. The heat loss from top-floor rooms to unused attic areas of a house or to unused attic peaks or eaves poses a heat loss problem. Louvers that are not closable generally are used in attics to permit the circulation of outside air. These louvers allow outside air in, even in the winter months. This air moves over the vapor barrier and insulation that encloses the upstairs rooms. The movement of this air carries away moisture that may have escaped through the vapor barrier from the room into the attic. If this moisture is not removed, it will condense and sometimes freeze on the inside of the exterior materials and possibly cause structural deterioration. Attic fans are also used to move the moist air outdoors.

Another source of heat loss is at the edges of concrete slabs. Unlike basement floors, which have a steady, slow rate of heat loss due to the constant temperature of the surrounding earth, concrete slabs have heat losses that are affected by the distance of their edges to the outside air. To prevent cold floors, rigid and moisture-resistant insulation should be incorporated for at least 2 ft along the edge of the slab. This should be done all around the perimeter of the edge exposed to the outdoor air. This method is shown in Fig. 6.1. In many installations this edge is also heated by the addition of an air duct that carries warm forced air. Therefore, edge losses must be calculated and factored into the overall heat loss of a building.

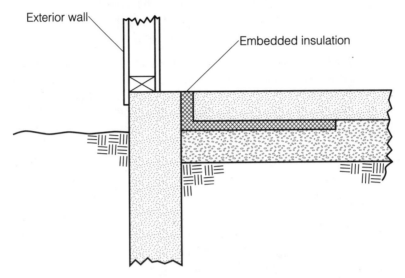

■ **6.1** *Insulation around the perimeter of the house's slab to reduce edge losses.*

Perhaps the greatest contributor to the heat loss in a home is infiltration. The winter winds blow cold air into indoor spaces through cracks around windows and doors on the windward side of the house. After this air has been warmed by the home's furnace, it tends to leave through similar cracks on the leeward side of the home. This happens because there is usually a reduced air pressure on the outside of the house on the leeward side. This reduced pressure creates a virtual suction, pulling the warmed air out through any cracks. Weatherstripping will greatly reduce the rate of air infiltration, but a house that is too airtight is also not recommended (because of moisture buildup, lack of fresh air, etc.). The best approach is to keep the air changes per hour to a minimum but have some for ventilation purposes.

Calculating Heat Loss

With any design of any house, whether existing or to be built, there will be a calculated value for the total heat loss of the structure. In addition, the structure must be evaluated for any heat gain. The heat gain, if any, must be factored into the overall heat loss to best size an efficient heating and cooling system. A heat-loss calculation is required to determine the heating system required. From here, the cooling system can be sized and the electrical requirements of the home can be predicted. All play a role in the energy efficiency of the entire structure. However, before we analyze the heat loss and heat gain of any structure, it is important to define the base and constants used. There are several coefficients that affect heat transmission. Their common letter designation and a description follow:

a is *air-space conductance* (Fig. 6.2). This is the Btu/h (British thermal units per hour) rate of heat flow through 1 ft² of area for a 1°F difference in temperature between the bounding surfaces. It is affected by position and by the emissivity E of the surfaces.

■ **6.2** *The air space between the block and brick has a heat conductance value* a. (Courtesy of Wayne J. Henchar Custom Homes.)

C is *conductance* through a material or combination of materials (Fig. 6.3). The British thermal unit-hour rate of heat flow through 1 ft² of a homogeneous material or a combination of materials for a 1°F difference in temperature between the exterior surfaces for the thickness of construction stated, not necessarily per inch of thickness.

E stands for *emissivity* (Fig. 6.4). This is the effective thermal emission (or absorption) of the surfaces bounding an air space.

f is the *film* or *surface conductance coefficient* (Fig. 6.5). This is the rate of heat flow in British thermal unit-hours through 1 ft² of surface due to the motion of air against the surface for a 1°F difference in tempera-

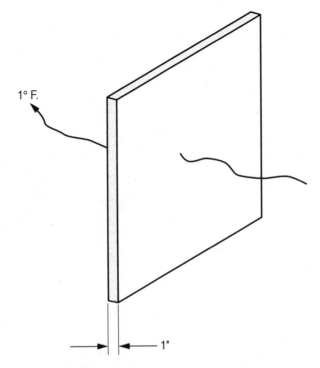

1° F.

1"

■ **6.3** *Heat transfer due to conductance* C.

ture. f_i is the symbol for the inside film coefficient, and f_o is the symbol for the outside film coefficient. These coefficients reflect the speed at which the air strikes the surface.

k is also *conductance* through a material, but for only a 1-in thickness (Fig. 6.6). This value is used in certain heat-loss calculations. It is expressed as the British thermal unit-hour rate of heat flow through 1 ft² of a 1-in-thick material for a 1°F difference in temperature between the two surfaces.

R is the *thermal resistance* of a material (Fig. 6.7). It is the reciprocal of heat transfer as expressed by such coefficients as U, C, f, or a. It may be expressed as hours per British thermal units for the standard 1 ft² of surface and 1°F temperature difference. For example, a wall with a U coefficient of 0.25 would have a thermal resistance R of 1/0.25 = 4.0. The R value is the most often used term in energy savings.

U is the *overall coefficient of heat transfer* (Fig. 6.8). The British thermal unit-hours flowing from air to air through 1 ft² of roof, wall, floor, or other building component in place in the structure under actual conditions for a difference of 1°F in temperature between the air on the inside and the air on the outside. This can apply to a combination of materials or to a single material such as glass.

Note: In energy-efficient building, it is important to strive for high R values and low C and k values. For example, a wall as shown in Fig. 6.9 has no

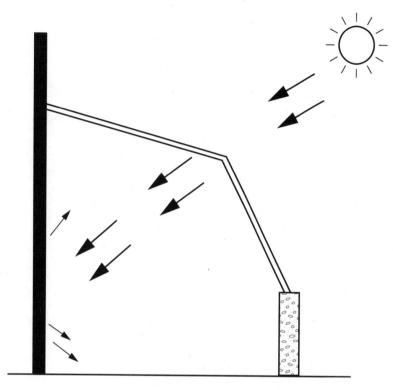

■ **6.4** *The effect of emissivity,* E, *on a dark-colored greenhouse wall.*

insulation. It has the following components: 4 in of clay brick, ½ in dead air space, ½ in of drywall, ⅜ in of sheathing, and building wrap. From Table 6.1 we find that the wall components have the following R values:

4 in clay brick	0.42
½ in dead air space	0.98
½ in drywall	0.42
Building wrap	0.17
⅜ in sheathing	2.22
Total R value	4.21

The U coefficient is equal to the reciprocal of the total R value:

$$U = \frac{1}{4.21} = 0.238$$

Therefore, if we have a wall 8 ft high by 15 ft long or 120 ft², and a temperature difference of, say, 70°F (inside temperature is 70°F, and outside temperature is 0°F), then our heat loss is going to be

$$120 \times 70 \times 0.238 = 1999 \text{ Btuh}$$

Glass, single pane, no film   Glass, single pane with film

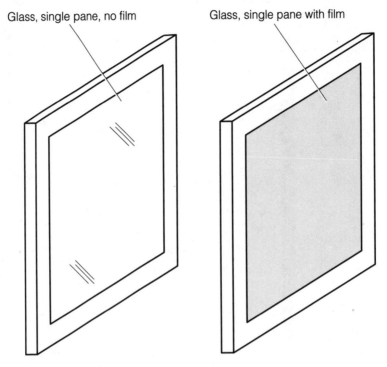

■ **6.5** *The effect of film conductance* f.

Now consider the wall section in Fig. 6.10, with 3½ in of fiberglass batt insulation. This is the same wall as was in Fig. 6.9 but with the addition of insulation. The *R* values of the components are as follows:

4 in clay brick	0.42
½ in dead air space	0.98
½ in drywall	0.42
Building wrap	0.17
⅜ in sheathing	2.22
3½ in insulation	10.92
Total *R* value	15.13

The *U* coefficient is equal to the reciprocal of the total *R* value:

$$U = \frac{1}{15.13} = 0.066$$

Therefore, the same wall, 8 ft high by 15 ft long, or 120 ft², with a temperature difference of, say, 70°F (inside temperature is 70°F, and outside temperature is 0°F) will have a heat loss of

$$120 \times 70 \times 0.066 = 554 \text{ Btuh}$$

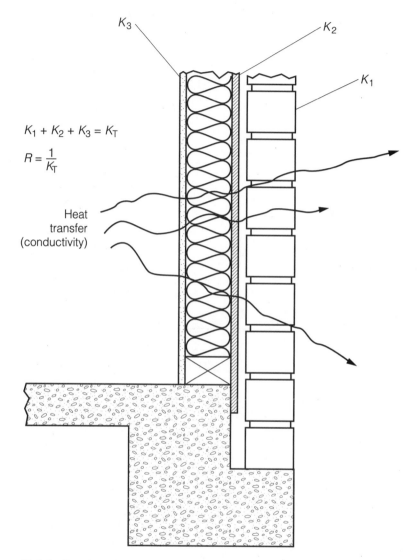

$$K_1 + K_2 + K_3 = K_T$$

$$R = \frac{1}{K_T}$$

Heat
transfer
(conductivity)

■ **6.6** *Conductivity* k_1 *the rate of heat through a ft² of material.*

Which wall is better for energy savings? Obviously, the section with insulation—by nearly a factor of 4 to 1. *Insulation is important!*

The R values in Table 6.1 are intended to be used as a guide and for the examples shown in this chapter. This list also does not contain every material used in the building construction industry. Extensive lists are available from home builders, most lumber yards, and some of the standards organizations listed in Chap. 5.

Figure 6.11 shows a multilevel house. It has four different levels, each with a unique total heat loss. Thus each level will have to be treated separately

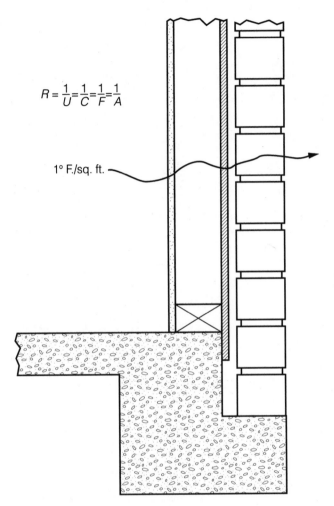

$$R = \frac{1}{U} = \frac{1}{C} = \frac{1}{F} = \frac{1}{A}$$

1° F./sq. ft.

■ **6.7** *Thermal resistance* R *through a typical wall section.*

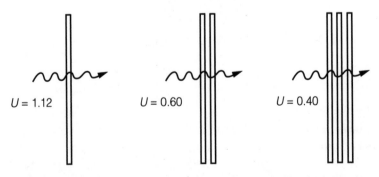

$U = 1.12$ $U = 0.60$ $U = 0.40$

■ **6.8** *Single, double, and triple panes of glass with air space between showing the overall coefficient of heat transfer* U.

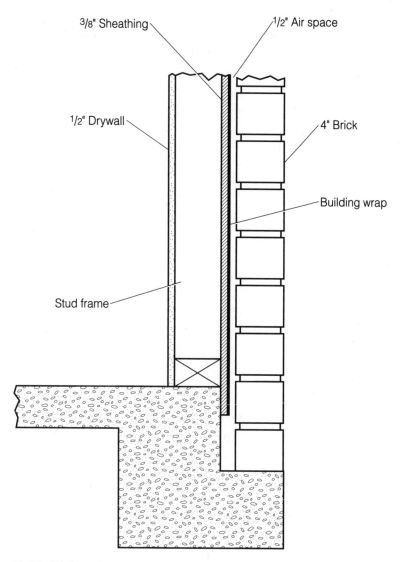

■ **6.9** *Wall section with no insulation.*

when calculating the entire house's total *hourly* heat loss. The design criteria are as follows:

The indoor temperature is 70°F.
The outdoor temperature is 5°F.
The design temperature difference is (70 – 5) 65°F.
The ground temperature constant is 55°F.
The wind velocity is 15 mi/h.

■ **Table 6.1** *R* **Values (Thermal Resistance) and**
U **Coefficients (Conductance) of Common Building Materials**

Building Material	R Value for Thickness Given	Conductance C for Thickness Given
Building wrap	0.17	5.88
⅜ in of sheathing	2.22	0.45
1 in of solid wood/plywood	1.25	0.80
8 in of concrete	1.49	0.67
8 in of concrete block	0.58	1.72
½ in of gypsum board, drywall	0.42	2.38
¾ in of interior wood finish	0.94	1.06
Wall materials		
Inside air film	0.69	1.45
½ in of dead air space	0.98	1.02
½ in of air space/reflective surface	2.66	0.38
1 in of fiberglass batt	3.12	0.32
1 in of expanded polystyrene	3.96	0.25
1 in of extruded polystyrene	4.96	0.20
1 in of rigid fiberglass	4.25	0.23
1 in of rigid urethane	7.50	0.13
½ in of hardboard siding	0.73	1.37
¾ in of drop siding	1.06	0.94
4 in of clay brick	0.42	2.38
⅝ in of stucco	0.12	8.33
Metal/vinyl siding with backing	1.41	0.71
Ceiling materials		
Inside air film	0.60	1.66
4 in of cellulose fiber	14.45	0.07
4 in of wood shavings	9.66	0.11
4 in of blown fiberglass	11.88	0.08
4 in of vermiculite	8.23	0.12
Asphalt shingles	0.45	2.22
Wood shingles	0.94	1.06
Floor materials		
Inside air film	0.93	1.07
Resilient floor coverings	0.08	12.5
Carpet (rubber/foam underlay)	1.29	0.78
Carpet (fiber underlay)	2.09	0.48
¾ in of hardwood flooring	0.69	1.45

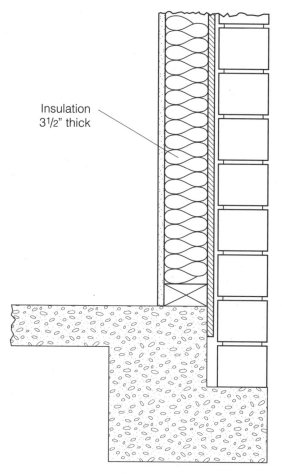

Insulation
3^1/$_2$" thick

■ **6.10** *Wall section with fiberglass insulation.*

■ **6.11** *Multilevel house example for heat-loss calculation.*

U coefficents (the reciprocals of the R values of the materials) to be used are as follows (note that all walls are exterior; ceilings that double as the floor for levels above do not have to be considered for heat loss, as well as any interior walls against heated spaces):

Building Surface	Materials	Total U Coefficient
Level 1 walls (above grade)	8 in of concrete block, ½ inch of air space, 4 in of brick	0.5
Level 1 walls (below grade)	8 in of concrete block	1.72
Level 2 walls (above grade)	8 in of concrete block, ½ in of air space, 4 in of brick, ½ in of drywall, 1 in of polystyrene, inside air film	0.142
Level 2 walls (below grade)	8 in of concrete block, ½ in of drywall, 1 in of polystyrene, inside air film	0.177
Level 3 walls (vinyl sides)	Vinyl siding, ⅜ in of sheathing, ½ in of drywall, 3½ in of fiberglass insulation	0.067
Level 3 walls (brick side)	4 in of brick, ⅜ in of sheathing, ½ in of drywall, 3½ in of fiberglass insulation	0.071
Level 4 walls	Vinyl siding, ⅜ in of sheathing, ½ in of drywall, 3½ in of fiberglass insulation	0.069
Roof	½ in of drywall, 6 in of fiberglass insulation, ½ in of plywood, inside air film, asphalt shingles	0.048

Now, let us compute the hourly heat loss for all the various levels of this multilevel house. The temperature to be used for the ground, to calculate under-grade temperature differentials, will be 55°F. Since the product of the constants for the density of air and the specific heat of air are 0.075 and 0.24, respectively, the value of 0.018 ($0.075 \times 0.24 = 0.018$) is used as the U value for air infiltration. There must be an infiltration value assigned for each level, a window and door heat loss value from Table 6.2, and an edge

■ **Table 6.2 Representative U Values**
for Doors, Windows, and Skylights

Description	U Value (Winter Ratings)
Plastic bubble skylights	
Single pane	1.15
Double pane	0.70
Windows	
Single glass	1.13
Double glass with	
½ in of air space	0.58

■ **Table 6.2 Representative *U* Values
for Doors, Windows, and Skylights (Continued)**

Description	*U* Value (Winter Ratings)
Triple glass with ½ in of air space	0.36
Doors	
1-in-thick solid wood	0.64
1.5-in-thick solid wood	0.49
2-in-thick solid wood	0.43

loss attributed to slabs on grade of 32 Btu per lineal foot. Each level will get a total Btu/h value assigned to it for use later in the whole-house loss tabulation.

Area Component	Area, Length, or Volume	*U* Coefficient	Temperature Differential	Btuh Loss
Level 1 (Fig. 6.12):				
Walls, above grade	364	0.5	65	11,830
Walls, below grade	100	1.72	10	1720
Doors (3)	144	0.55	65	5148
Slab edge loss	51 linear feet	32 Btu/ft		1632
Infiltration	5200 ft³/h	0.018	65	6552
Total heat loss for level 1				26,882
Level 2 (Fig. 6.13):				
Walls, above grade	306	0.142	65	2824
Walls, below grade	188	0.177	10	333
Doors (1, sliding glass)	42	0.70	65	1911
Windows (3)	27	0.66	65	1158
Slab edge loss	25 linear feet	32 Btu/ft		800
Infiltration	5250 ft³/h	0.018	65	6142
Total heat loss for level 2				13,168
Level 3 (Fig. 6.14):				
Walls, vinyl	357	0.067	65	1555
Walls, brick	181	0.071	65	788
Doors (2 w/glass)	63	0.70	65	2866
Windows (5)	57	0.66	65	2445
Infiltration	7290 ft³/h	0.018	65	8529
Roof	729	0.048	65	2274
Total heat loss for level 3				18,457
Level 4 (Fig. 6.15):				
Walls	559	0.067	65	2434

Windows (5)	57	0.66	65	2445
Skylights (1)	5	0.70	65	228
Infiltration	5568 ft³/h	0.018	65	6514
Roof	691	0.048	65	2156
Total heat loss for level 4				13,777

Heat Loss Summary
for Entire House

Level	Btuh
Level 1	26,882
Level 2	13,168
Level 3	18,457
Level 4	13,777
Total heat loss	72,284

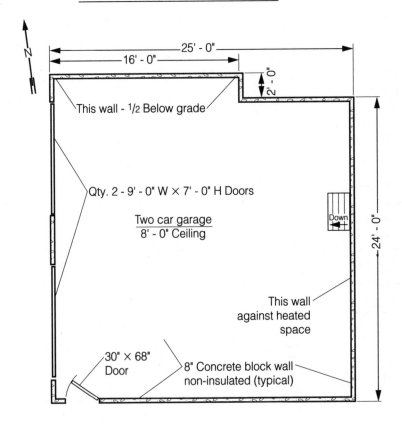

Level one

■ **6.12** *Level 1 of a multilevel house, floor plan.*

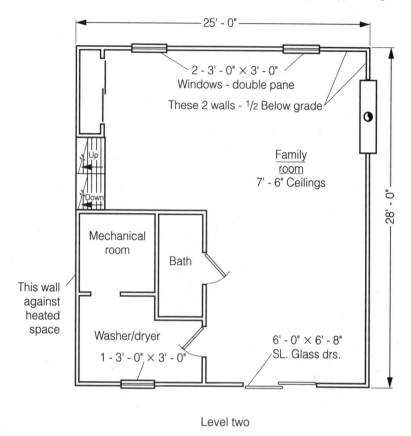

Level two

■ **6.13** *Level 2 of a multilevel house, floor plan.*

Whenever you do a heat-loss calculation, some common sense has to be used. For instance, in an existing home, you may not know exactly what materials or thicknesses of materials were used. Therefore, you may have to make various assumptions or "educated guesses" in order to do a heat-loss calculation. Today, a wide variety of software programs is available that will allow you to run many different combinations of materials and thicknesses. Running the software program several times certainly beats tearing apart a wall to see what's there.

What the heat-loss program and calculations show the home owner and home builder is where the worst areas of the house are in terms of heat loss. Obviously, there has to be some heat loss. Some houses are better insulated and sealed better than others. However, take a good look at the heat-loss calculations and see where the areas of highest heat loss are. This directs the effort toward further insulating or remedying of the situation.

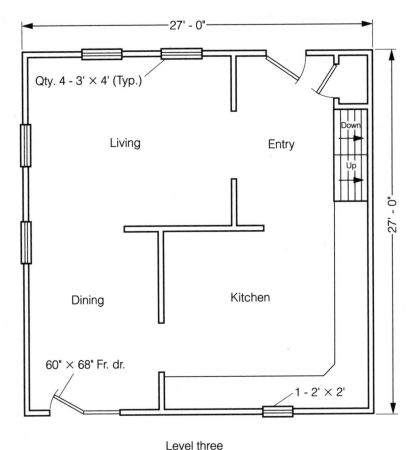

Level three

■ **6.14** *Level 3 of a multilevel house, floor plan.*

Thus the heating or cooling system should have a capacity of no more than 10 percent in excess of the calculated requirement. Many years ago, when energy was not too much of a concern, heating equipment often was oversized, sometimes by a factor of two to three times. Installing the smallest-capacity heating equipment to meet the load will save both energy and money. Calculation of total home heat loss is generally done by the mechanical heating contractor but can be done by the home builder. Contractors who are inexperienced in low-energy house design and heat loss may still oversize the heating system (it's not their energy dollars being wasted years down the road). Therefore, it may be in the home owner's best interest to either verify the sizing by performing his or her own heat-loss calculation or by having a second opinion from another contractor. Once the equipment is installed, it is obviously too late for improvement.

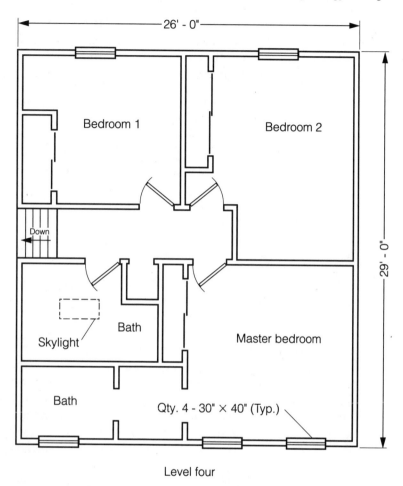

Level four

■ **6.15** *Level 4 of a multilevel house, floor plan.*

As was shown earlier in this chapter, calculating the heat loss from a home is quite simple. The heating requirement will be highest when the outside temperature is lowest and there is no solar gain. A cold, windy winter night is when the heating load will be greatest. Heat flows out through all the building surfaces, including walls, ceilings, floors, windows, and doors. We are striving for a temperature balance point, a point where the outside air temperature and the inside air temperature of a particular building will be in thermal equilibrium, requiring neither heating nor cooling to support thermal comfort. Because this only occurs a few times during the year (depending on the region of the country in which one lives), design temperature is at the predicted worst. In the preceding heat-loss example it was

5°F outside, and we desired 70°F inside. Thus our temperature differential was 65°F. Design temperatures and conditions are available for most parts of the country. Many utilities have this information for a given area, or it can be obtained from the home builder or mechanical contractor.

We are also trying to design out of the residence any temperature swings. Temperature swings are the variations in temperature within a room or home and are measured in degrees. People can sense more than 2°F temperature swings or else they are not comfortable. Cold, windy wintry days are the biggest challenge to minimizing temperature swings. Heat-loss calculations usually are based on the amount of heating and cooling degree days a region will see in a given year. A *heating degree day* is defined as follows: Each degree the average temperature for a day is below 65°F produces one heating degree day. *Cooling degree days* are calculated for each degree that the average temperature is above 65°F.

During sunny days, there is actually some heat gain in most homes, especially energy-efficient homes. There are also a number of sources of heat gain in a typical house. Not only do the occupants give off heat, appliances and lights contribute significantly to home heating. Each person can provide about 75 W of heating energy, whereas 200 or 300 W is available from appliances such as freezers, ranges, refrigerators, and so on. The average home therefore provides 500 W or more daily of the total energy required for space heating. However, when calculating the heat loss and sizing for a heating system, heat gain must be ignored because the worst conditions have to be considered or else the heat will not be able to be delivered when the home most needs it. Heat-gain systems are analyzed further in Chap. 11.

A large number of computer software programs are available that can be used to more accurately calculate building heat loss. These programs require a detailed breakdown of each building component and complete area weather data. Most of the programs available require a considerable learning curve and often are not practical unless you do a lot of heat-loss calculations, are a house designer, or are designing a complex solar building.

Heating System Sizing

The efficiency of the heat source must be taken into account when selecting a heating system. The house in Fig. 6.11 was a fairly large house, with nearly 3000 ft² of usable heating space. In the example, a 21-kW (divide the total Btu/h by 3.415 to get kilowatts) heat source would be needed (72,000 Btuh). However, if the two-car garage were not to be heated, then almost 27,000 Btuh can be taken from the calculations and sizing. A much smaller heating system could be installed (45,000 Btuh), and frequently, garages are *not* heated. A garage is a drafty area, and the heat loss is typically the worst in the entire residence. If one chooses a 100 percent efficient electric heating source, the exact figure calculated above can be used to size equipment. Gas furnaces range from 70 to 80 percent efficient and are measured and rated seasonally over an entire year of operation (SEER rating). Divide the heat load (21 kW) by the system efficiency (0.70) to obtain the bonnet size of 30 kW (102,000 Btuh) necessary to provide 21 kW. Gas-fired furnace and boiler units with efficiencies of 90 to 95 percent are also available but usually are produced in large output sizes and are more expensive. Thus, in this example, the home owner would have two decisions to make:

1. Does the garage get heated? If not, then the size of the furnace can be much smaller, by nearly 40 percent.

2. Will a higher-efficiency furnace be purchased? If so, then the bonnet size can be smaller yet (21 kW divided by 0.95 = 22 kW) and thus the furnace smaller.

To get the most efficiency with this house, the garage probably should not be heated, and the furnace purchased should be a high-efficiency unit. This would be the optimal scenario. However, there are still areas within the house that can be upgraded to make it even more efficient. Use the heat-loss calculation worksheet (Fig. 6.16) to pinpoint those areas of inefficiency and correct them.

Common Household Appliances and Their Contribution (or Lack Thereof) to the Residence's Energy Efficiency

All the appliances within a home could be called the *appliance system*. It is a system because there are so many different devices and together they represent virtually all the electrical load within the house. They can be grouped in several categories: kitchen appliances, lighting devices, home entertainment, laundry, hot water, bedroom/bath, lawn/garden, and the big one—heating/cooling. Table 6.3 shows several electrical appliances or pieces of equipment that may be found in the typical residence. It is surprising to see such a long list. No wonder we use the most electric power per capita in the world. In Table 6.3, each appliance is listed with its corresponding wattage. Obviously, electricity is only consumed whenever the appliance is "on." The table shows many appliances, and if each were turned on at the same time, the electricity draw would be so great that the main house circuit breaker would trip. More amazing is the fact that Table 6.3 is still far short of all the possible appliances available for purchase, not to mention new ones yet to emerge. We do love our electrical-based gadgets!

Heat loss calculation worksheet for residential construction.
(Area times U value times temperature difference, inside to outside
= Heat loss)

Area of house	Heat loss source	xArea, length or volume	xU value or other	xTemperature difference (I-O)	=Heat loss in btuh
1.	Walls Glass Doors Roof Infiltration Edge loss				
Subtotal					
2.	Walls Glass Doors Roof Infiltration Edge loss				
Subtotal					
3.	Walls Glass Doors Roof Infiltration Edge loss				
Subtotal					
4	Walls Glass Doors Roof Infiltration Edge loss				
Subtotal					
Summary					
Total heat loss					

■ **6.16** *Heat-loss calculation worksheet.*

■ **Table 6.3 Electrical Equipment and Appliances, with Wattage Rating**

Appliance	Wattage
Kitchen	
Blender	720
Broiler	1200
Can opener	100
Carving knife	100
Coffee maker	
Automatic percolator	850, brew
	100, warm
Automatic drip	1500, brew
	100, warm
Convection oven	1500
Corn popper	
Oil	575
Hot air	1400
Deep fat fryer	1500
Dishwasher	1200
(*not* including hot water)	
Garbage disposal	420
Fondue/chafing dish*	800
Food dehydrator*	875
Food processor	690
Fry pan*	1200
Griddle*	1200
Mixer	
Hand	120
Standing	150
Ice cream freezer	130
Juicer	90
Knife sharpener	100
Microwave oven	
Full size	1500
Compact	1000
Pressure cooker	1300
Range, self-cleaning oven*	3000
Slow cooker	
Low	75
High	150
Toaster	1100
Toaster oven	1400
Trash compactor	400
Waffle iron*	1200
Wok pan*	1000

■ Table 6.3 Electrical Equipment and Appliances, with Wattage Rating (Continued)

Appliance	Wattage
Audio/visual	
Clock radio	4
Compact disc player	12
Computer, personal	150
Copy machine	525
Electric train	15
Movie projector	150
Computer printer	120
Slide projector	250
Stereo	125
Television, color	
Tube type	240
Solid state	100
Big screen	360
Television, black and white	
Tube type	100
Solid state	55
Videocassette recorder	50
Video games	9
Laundry	
Clothes dryer*	5000
Iron*	1100
Washer	
Automatic	500
Wringer type	280
Water softener	4
Bedroom and bath	
Blanket*	175
Curling iron	40
Hair dryer	
Soft bonnet	400
Hard bonnet	1200
Hand held	1200
Hair rollers	350
Heating pad*	65
Heat lamp	250
Makeup mirror	25
Massager	15
Shaver	14
Shaving cream dispenser	60
Sun lamp	250
Toothbrush	7

Appliance	Wattage
Bedrooms and bath	
Vaporizer	480
Waterbed heater*	400
(covered bed, in a heated room)	
Lawn and garden	
Edger	480
Grass trimmer	240
Hedge trimmer	288
Lawn mower	1200
Chain saw	1380
Circular saw	1150
Electric drill	287
Jig saw	287
Sander	287
Table saw	1380
Electric fence	10
Heat tape*	100
Leaf blower	720
Pool filter	750
Saw (band)	960
Shop vacuum	825
Snow blower	1200
Welder	2000
Lighting	
Bug light	25
Candles, decorative	50
(10 @ 5 W)	
Christmas tree	
64 lights (5 W)	320
Miniature (50 bulbs)	18
Fluorescent light	34
Pathway lights	
(low-voltage, 6 @ 18 W)	108
Post light	
Incandescent	100
Mercury	50
Quartz, spot or flood	300
Security light	
Mercury	175
High-pressure sodium	70
Table lamp	75
Yard light (flood or spot)	100
Light bulb	100

■ **Table 6.3 Electrical Equipment and
Appliances, with Wattage Rating (Continued)**

Appliance	Wattage
Miscellaneous	
Appliance trimmer	2
Battery rechargers	10
Central vacuum system	1000
Clock	2
Cordless appliances	5
Fish tank	
Heater*	25
Aerator	3
Floor polisher	400
Garage door opener	350
Incinerator	605
Kiln, ceramic*	6400
Outdoor grill	1800
Septic tank aerator	500
Sewing machine	75
Sump pump	500
Vacuum cleaner	650
Water pump	500

*Controlled by thermostat.

Home Energy Rating System

The Home Energy Rating System (HERS) measures and rates the relative energy efficiency of any house, regardless of its age, location, construction type, or fuel use. The rating evaluates the performance of the thermal envelope, glazing strategies, siting and orientation, the heating, ventilation, and air conditioning (HVAC) system, and other criteria and is obtained by an onsite inspection. HERS calculations include estimates of annual energy performance and costs, and can provide insight into cost-effective, energy-efficiency improvements. The HERS provides objective, standardized information on the energy performance of homes. Expanded use of the HERS could stimulate use of energy-efficiency technologies by making energy efficiency a more quantifiable, visible, and recognized attribute as homes are designed, built, bought or sold, and improved.

Energy-efficient-financed (EEF) products recognize the energy cost savings of energy-efficient homes and enable borrowers to increase their buying power by providing the longer-term lower interest rates and tax benefits of mortgage financing. The small increase in the mortgage payment as a result of an EEF improvement is typically more than offset by the energy savings. Through this program, products also can be financed for energy-saving improvements to existing homes as part of the mortgage at the time of purchase or refinancing. An energy-efficient loan permits more leniency in qualifying borrowers for mortgages on homes that perform at a higher level of energy efficiency. To receive information about the HERS or to request any software or publications, please refer to Chap. 5 for contacts and addresses.

Trickle Currents (Power Losses Inside the Home)

We have all probably noticed by now that some of our appliances seem to have various lights or displays still "on" even when we turn the appliance "off." The fact that these lights and displays are still "on" means that some power or electricity is still flowing through the device. For example, the digital clocks on videocassette recorders (VCRs) function just fine even with the power supposedly shut off. More and more, any device that has a clock-based microprocessor in it has got to be receiving some amount of electric current in order to keep updating. As we go forward, there will be other functions that manufacturers may feel need to be powered, even when the main appliance switch is turned "off" (usually by a battery-powered remote controller). If they incorporate other "on when its supposed to be off" functions, more electricity will flow, and thus there will be more "leaking current." This current is leaking to us but controlled slowly by the electronic system. Also, if more and more appliances have to have these "always on" functions (watch out for any preheat, trickle, or keep warm functions, they will be heavy users of electric current), then the overall leaking load will increase (so will your electric bill).

Home owners or home buyers beware! Not many could pinpoint every leaking appliance within their very own residence. However, wouldn't it be nice to know where electric current is flowing when we go to bed at night? Sure, we could pull all the plugs to every appliance, but that's not too practical. Why can't the manufacturers tell us how much trickle current and watts are required to have an appliance work at best performance. Then we could add up the total losses. After all, the Food and Drug Administration makes all types of labels be applied to food, why not with televisions, microwave ovens, VCRs, and the stereo equipment? As a project, count all the various appliances and devices within your own home that have to have electricity still passing through them, even though they are supposed to be off. The television, VCR, stereo, microwave, electric range, printer, computer, and coffee maker are included, and the list goes on.

The amount of electricity used every month by these "leaking" or "trickle" currents *does* add up. Depressing the power switch may blacken the television screen, stop the music on the stereo, or stop the heating element on the coffee maker, but these appliances continue to draw power. They need it to maintain memory of certain settings, to keep a remote-control sensor alert, or for a variety of other functions deemed important by the manufacturer. Many newer computers, cable television boxes, VCRs, ceiling fans, cordless drills, and video games likewise stay on even though switched off. And just how many wall power packs do you have plugged in at your house? Figure 6.17 shows the wall pack in the appliance or in the wall, either way it is still drawing current.

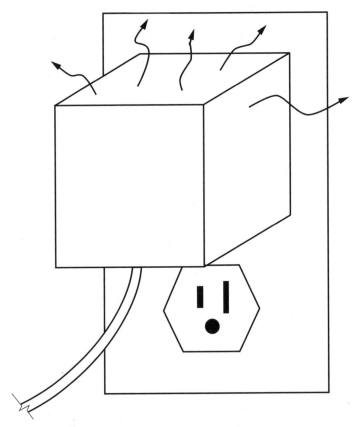

■ **6.17** *Common household wall power pack (heat losses).*

The wall pack is another energy absorber that is robbing electricity from your home practically around the clock. The wall power pack is another name for a power supply or voltage matching transformer that is needed for many of our electronic devices. Power packs also tend to get warm, and warmth is a sign of wasted electricity. As another project, count all the wall power packs in your house. They are usually found by the remote telephones, answering machines, printers, facsimile machines, and so on.

Thus, if the total value of watts from all the trickle currents flowing through many household appliances was to be added to the power used up each month by the various power packs, then an even greater portion of the electric bill is basically, wasted energy. All this to make our lives more immediate and convenient. Is this really necessary?

A typical home may use between 75 and 100 W to power devices that supposedly are not on, but are still drawing electrical current. They may have a red or green light-emitting diode (LED) energized, and thus we know there is some electricity flowing into them. However, there are many appliances and pieces of equipment that deceive us: We think they're off, and they're not. Such unnoticed consumption, commonly referred to as *leaking current*, constitutes the electric analog of heat seeping out of poorly insulated homes. If we are striving to better insulate our homes and save energy, then shouldn't this issue be addressed? In some respects, it is happening with the Energy Star program. A new program will extend EPA's Energy Star program to televisions and VCRs. For example, any television that consumes 3 W or less in its off or standby mode will earn the right to identify this feature by carrying the Energy Star label on its package as well as its advertising.

This leaking electricity may not seem like a large amount to anyone, individual household (and it isn't), but multiply the amount by millions of households, and all of a sudden this is very costly, perhaps in the billions of dollars. Besides the wasted money, there is the issue of pollution and the environment. To supply the estimated amount of electricity demanded by leaking household appliances worldwide, power plants are already sending several million tons of carbon and soot into the atmosphere every year. The risk of global warming only gets more dramatic. Several organizations are already trying to raise public awareness of the problem and what can be done to limit it. A few also will target manufacturers and policy makers. Most electricity leaks trace to power transformers (power packs) that cannot be turned off. These power transformers are an electric cord attached to a small black-box transformer that plugs into the wall. These units allow battery-powered radios, tape recorders, and other appliances to run off household current. The typical home may have anywhere from 2 to 12 wall pack units plugged in, each stepping the residential wiring's high-voltage alternating current (120 vac) down to the lower, direct current (3, 9, or 12 vdc) needed by some appliances. These losses all add up!

Electricity leaks also stem from a manufacturer's desire to make their products more versatile and more appealing to the consumer without regard for leaking current. In televisions and VCRs, for instance, the leaking current may power a clock, a digital display, and a memory chip that recalls what station the television was last tuned to. In portable phones, rechargeable flashlights, and hand-held vacuum cleaners, it may continuously power a battery charger—long after the appliance is fully charged. For products sold with remote controls, it really only powers the sensor that awaits the remote's call. This category of electrical leaks arises because these appliances remain in a standby mode when switched off. Even though this state of readiness may draw only a fraction of the electricity consumed when the product is in active use, the leaking kilowatt hours can add up. There are also some devices whose power draw never varies, such as the satellite receivers that, like cable boxes, can expand the range of stations available to a television. Many of these devices consume roughly the same amount of power, on or off. The only difference seems to be whether the red LED "on" light is powered.

The VCR industry had said it would not be possible to go below 6 W in standby, but already many are now at 1 W. Thus it can be done. Today with appliances, except for most kitchen appliances, there is a growing popularity to use the remote control. When consumer electronics are shut down with remotes, they do not turn off completely; rather, they enter a dormant standby that allows them to remain sensitive to the remote's command. One solution is an actual hard "off" switch for the device that kills all electricity to the unit. Many countries in Europe already use this standard.

More efficient wall power packs have to be built. "Intelligent" wall packs also can help. A tiny integrated circuit chip in the wall pack and another in the appliance's power switch would result in a "smart" transformer that would supply current if and only if it sensed an appliance had been turned on. To make the change without adding too much to the cost of generic wall power packs is the challenge. In this way, the transformer goes to sleep whenever the product to which it is attached is turned off. It wakes up at programmed intervals, from a fraction of a second up to an hour to see if the appliance is back on. If so, the transformer supplies it with power. If not, it dozes off again. Though this transformer does not plug into a wall, there is no reason, in principle, why it could not be reconfigured as a wall pack. It would probably add about $25 to the cost of an appliance (phones, answering machines, etc.). However, it is too costly for products retailing under $50, as the smart wall pack would make up 50% of the cost of the appliance.

Many appliances, such as televisions, computers, and VCRs, can significantly limit their standby power consumption if their transformers employ "switch mode" technology. This technology essentially switches the power to a device on and off every 10 millisec, sensing on each cycle whether power is required and supplying it only when necessary. This allows you to control how much power is delivered from your wall socket to the appliance, keeping leakage rates low. Another possible solution is a rechargeable backup battery in VCRs and other products that need to keep some memory chips or other low-power devices energized. The battery could be recharged by the wall current whenever the appliance is in active use.

Energy Tracking and Load Monitoring

There are many devices on the market today that can perform data logging and monitoring of residential electric circuits. Many automatic meter readers can monitor gas and water use also. As long as it is a pulse-type metering application, then it can be monitored. With a maximum pulse rate capability of 1 kHz, these energy-monitoring systems can measure the pulse from a standard meter. Power monitoring along with load profiling for all energy fuels and their use and demand can be attained. As we move into the twenty-first century, the need to monitor and tie into home automation systems will increase. Another type, called a *nonintrusive appliance load monitor* (NALM), is designed to monitor an electric circuit that contains a number of appliances. Typically, these devices switch on and off independently. By a sophisticated analysis of the current and voltage waveforms of the total load, the number and nature of the individual loads, their individual energy consumption, and other relevant statistics such as time-of-day variations are monitored. No access to the individual components is necessary for installing sensors or making measurements. This can provide a very convenient and effective method of gathering load data compared with traditional means of placing sensors on each of the individual components of the load. The resulting end-use load data are extremely valuable to consumers, energy auditors, utilities, public policymakers, and appliance manufacturers for a broad range of purposes. For example, a monitor placed outside a home can determine how much energy goes into each of the major appliances within the home.

The appliance load monitor analyzes the total load, checking for certain signatures that provide information about the activity of the appliances that constitute the load. For example, if the residence contains a refrigerator that consumes 250 W, then a step increase of that characteristic size indicates that the refrigerator turned on, and a decrease of that size indicates the turn-off events. Other appliances have other characteristic signatures. After determining the exact on and off times from the signature events, any desired statistics, such as energy consumption versus time of day or temperature, can be tabulated. Knowing the time of each on and off event, the total energy consumption of the refrigerator and the heater are easily determined. By also considering measurements of the total reactive power or harmonic current, along with the real power shown, changes in the resulting vector function of time would reveal even more information about the particular appliances. Traditional load research instrumentation involves complex data-gathering hardware but simple software. A monitoring point at each appliance of interest and wires connecting each to a central data-gathering location provide separate data paths, so the software merely has to tabulate the data arriving over these separate hardware channels. The nonintrusive approach reverses this balance, with simple hardware but complex software for signal processing and analysis. Only a single point in the circuit uses instrumentation, whereas mathematical algorithms separate the measured load into its different components. In many load-monitoring applications, this is a very cost-effective tradeoff, which is a major advantage of a nonintrusive appliance load monitoring system.

Appliances can be broken down into various groups. They are (1) the on/off, or two-state, appliances, such as light bulbs or toasters, which are either on or off at any given moment (early research focused on techniques for monitoring these), (2) multistate appliances, such as washing machines or dishwashers, with distinct types of on states, e.g., fill, rinse, spin, pump, etc. (recent research has extended the methods to apply to the multistate case), and (3) continuously variable appliances, such as light dimmers and variable-speed hand tools, with a continuous range of on states (these are difficult to monitor because they do not generate step changes in power; they are sometimes referred to as *nonlinear loads*). To monitor any of the types of appliances, a personal computer–based system collects samples of real and reactive power consumption over time and automatically learns the control structure of the load in real time, drawing its finite-state diagram. The system also reports the load's state at each point in time, the total time spent in each state, and total energy consumption for each of the states.

Closing Remarks on Measuring Energy Savings

The home owner cannot have energy savings unless some energy is being wasted. With the thought that any device, no matter how efficient it is, can be made more efficient, then there can always be some amount of energy savings to strive for. However, the home owner and home builder have to know where to look for any instances of energy loss, and they also need to know how much is being lost; they have to have this point of reference. Thus we all need to know how to measure energy savings. Whether it is in lost British thermal units through the wall and out to the cold winter winds or in leaking electricity in small 10- to 20-W amounts, there can always be some betterment to these losses of energy.

With more and more technology becoming available, both hardware and software, knowing where every watt and British thermal unit goes will become commonplace. Perhaps these advancements will only help to make energy savings more automatic. Human nature is to customarily talk about an issue but let someone else take care of it. This does not always hold true whenever the home owner has to pay higher rates for energy. Such a home owner will try to lessen the energy burden on his or her residence. One fact we can present as certain: Energy costs will always continue to increase. Some people even think that energy sources drive the costs of most other goods and products we use. Finding every way to save a watt or a British thermal unit will pay off in the long run.

Finishing Systems and Energy-Saving Products

7

While it is important to gain energy savings with every type of material used, there does not necessarily have to be a tradeoff between decorative and energy-efficient materials. As was seen in Chap. 6, virtually all building materials have some thermal resistance. While some are much better insulators than others, each layer of material contributes. When it comes time to finish a home, these last materials are not called on to provide dramatic increases in energy savings but rather are the touches to make everything look aesthetic and attractive. However, these final coatings can make some contribution to the heat gain of a house and thus contribute to the energy-savings bottom line.

Additionally, the physical location of the doors and windows within rooms may have to depart from the traditional. Frequently, floor plans take on the same look without regard to the overall orientation of the house with respect to the sun. During the design stage, especially with some computer-aided design (CAD) systems and modeling, experimentation with the location of various built-in components may be worthwhile. Proper placement of windows and doors, skylights, fans, and other building elements that have some direct affect on the home's energy savings does not really increase the cost of the structure whatsoever. Instead of installing a larger, triple-glazed window unit (obviously more expensive), why not move the double-glazed window unit to a more optimal location within the room. This should not increase the costs at all.

The home builder becomes the designer, offering architectural and aesthetic renderings of the energy-efficient residence to the prospective home owner. True architecture is making the structure look great while providing function. This is one of the most important jobs of the home builder. Architects are trained to provide this in their works but are rarely involved with the majority of home designs and actual construction. This is where the home builder must wear his or her "architect's hat" and listen to the needs and desires of the home owner, all the while keeping the design energy efficient. For instance, dormers are often requested by home owners because they provide a nice look to the outside of the house and add a little extra space to the upper floor rooms. However, because a dormer is an actual break in the roof's uniform construction, there is a better probability for leaks, air infiltration, and heat loss. Additionally, the cost of constructing dormers escalates the overall price of the house.

When discussing finishing systems for a residence and their contribution to energy efficiency, one should focus on a handful of factors that make a difference. The colors chosen for the roofs, exterior walls, interior walls, and floors and the surface textures are extremely important. Likewise, the material types and their locations with respect to the sun and the building's orientation make a difference. Other finishing systems include tapes, sealants, and caulking—all used to keep infiltration, air entering through building cracks, and exfiltration, air exiting through building cracks, to a minimum. Tinting and the use of film at windows help reduce cooling costs by letting light in but keeping heat and glare out. Many other finishing products can make some kind of contribution, and the home owner and home builder should not neglect this fact.

Emissivity, Absorption, and Material Colors

All building materials can be used in accordance with their ability to absorb or reflect radiant energy. Radiant heat transfer differs from both conduction and convection in that a medium is not required to transfer the heat. Radiant heat transfer is an electromagnetic phenomenon similar to light transmission, radio waves, and x-rays. A net exchange of heat occurs when the absorption of radiant energy by one body exceeds the energy that it is radiating. A body that absorbs all the radiation that strikes it is referred to as a *black body*. A black body has an emittance or emissivity value of 1. All materials reflect as well as absorb thermal radiation. Therefore, the fraction of the heat that is reflected is called the *reflectivity* of the material, the fraction absorbed is called the *absorptivity*, and the effectiveness of the material as a thermal radiator at a given temperature is known as its *emissivity*.

Emissivity has a ratio of the emission of heat at a given temperature to the emission of heat from a perfect black body at the same temperature. The equation looks like this:

$$E = \frac{R_M}{R_B} = \frac{\text{radiation from material}}{\text{radiation from black body}}$$

With the sun being a tremendous radiator of heat, a house can be a heat gain whenever and wherever the home builder decides. The emissivity of materials is related to their interaction with radiant energy or electromagnetic radiation such as solar energy and long-wavelength infrared energy radiation from warm objects. Convection and conduction modes of heat energy transfer are not affected by the emissivity characteristics of materials. As mentioned earlier, emissivity is inversely related to the infrared radiant energy reflectivity of materials. The higher the reflectivity, the lower is the emissivity. Highly reflective materials such as the polished surfaces of silver, gold, or copper have low emissivities. Low-*E* coated glass, while transparent to visible light, is highly reflective to long-wave infrared energy and therefore exhibits a low emissivity.

Table 7.1 shows various materials and their emissivity values. Since the emissivity of a material will vary as a function of temperature and surface

■ **Table 7.1 Emissivity *E* of Materials**

Material	Emissivity
Adobe	0.90
Cement	0.96
Cement, red	0.67
Cement, white	0.65
Cloth	0.90
Paper	0.93
Slate	0.72
Asphalt, pavement	0.93
Asphalt, tar paper	0.93
Basalt	0.72
Brick, red, rough	0.93
Fire clay	0.75
Lime clay	0.43
Fire brick	0.75–0.80
Gray brick	0.75
Silica, glazed	0.88
Silica, unglazed	0.80
Sandlime	0.59–0.63
Carborundum	0.92
Earthenware, glazed	0.90
Earthenware, matte	0.93
Porcelain	0.92
Clay	0.39
Clay, fired	0.91
Clay, shale	0.69
Clay tiles, light red	0.32–0.34

■ **Table 7.1 Emissivity**
***E* of Materials (Continued)**

Material	Emissivity
Clay tiles, dark purple	0.78
Concrete, rough	0.94
Tiles, natural	0.63–0.62
Tile, brown	0.87–0.83
Tile, black	0.94–0.91
Cotton cloth	0.77
Dolomite lime	0.41
Glass, smooth	0.92–0.94
Granite	0.45
Gravel	0.28
Gypsum	0.80–0.90
Ice, smooth	0.97
Ice, rough	0.98
Lime mortar	0.90–0.92
Limestone	0.95
Marble, white	0.95
Marble, smooth, white	0.56
Marble, polished, gray	0.75
Mica	0.75
Linseed oil, on uncoated aluminum foil	0.09
Linseed oil, on two coats aluminum foil	0.56
Paint, blue	0.94
Paint, black	0.96
Paint, green	0.92
Paint, red	0.91
Paint, yellow	0.90
Paint, aluminum	0.27–0.67
Paint, bronze	0.34–0.80
Quartz, rough, fused	0.93
Rubber, hard	0.94
Rubber, soft, gray	0.86
Sand	0.76
Sandstone	0.67
Sandstone, red	0.60–0.83
Sawdust	0.75
Shale	0.69
Silica, glazed	0.85
Silica, unglazed	0.75
Silicon carbide	0.83–0.96
Silk cloth	0.78
Slate	0.67–0.80
Snow, fine particles	0.82

Material	Emissivity
Snow, granular	0.89
Soil, surface	0.38
Soil, black loam	0.66
Soil, plowed field	0.38
Coal	0.95
Stonework	0.67
Water	0.67
Waterglass	0.96
Wood	0.80–0.90
Wood, beech, planed	0.94
Wood, oak, planed	0.91

finish, the values in this table should be used only as a guide for relative degrees of measurement between materials. The exact emissivity of a material should be determined when absolute measurements are required. Additionally, Table 7.2 shows relative absorption values for several com-

■ Table 7.2 Absorption Properties of Colored Materials

Material	Absorption Value
Aluminum foil	0.15
Black tar paper	0.93
Concrete	0.60
Dry sand	0.82
Flat back paint	0.96
Fresh snow	0.13
Galvanized steel	0.65
Granite	0.55
Graphite	0.78
Green paint	0.50
Green roll roofing	0.88
Gray paint	0.75
Red brick	0.55
Red paint	0.74
Water	0.94
White enamel	0.35
White paint	0.20
White plaster	0.07

NOTE: An absorption value of 1.00 represents 100 percent absorption.

mon materials. As a comparison, a flat black paint has an absorption value of 96 percent, whereas a white paint absorbs only 20 percent. Thus, whenever selecting a particular color for a portion of a house, it should be determined if that area of the house should absorb heat or reflect it. This is true for roofs and exterior walls. In sunny southern climates, it is necessary to keep these materials a lighter color to help reduce cooling loads. Conversely, in northern climates, the colorization should be darker to help with solar heat gains. Maybe with the advancement of home automation systems will come the changing of color schemes for exterior walls and roofs, with lighter shades in the summer and darker shades in the winter. Much like the chameleon, the house should be able to change the color of its skin without the owner having to repaint each season. At any rate, this is an important area for the home builder and home owner to discuss and agree on, and the foresight and time spent reviewing this phenomenon can pay back energy savings for many years.

Finishing Materials

Traditional methods of finishing houses call for the use of drywall, lath and plaster, and conventional floor tiles. Whenever practical, materials of higher R values can be substituted. For instance, wood board can be used instead of drywall for many walls. From the R value table in Chap. 6, we have seen that the thermal resistance value of wood is better than that of drywall or gypsum. Obviously, there may be a cost premium to evaluate, but wood finishing products and wood substitutes are very attractive as well as energy efficient. Additionally, masonite and medium-density fiberboards are used often to cover wood studs, especially in retrofit applications. Lumber yards or home centers can provide the R values of the products they sell. It is always wise to ask for this value and compare different materials and different thicknesses of multiple types of these 4×8 ft boards. Then the most energy-efficient package for the money (best value) can be selected. This is very important whenever the walls to which the board will be applied are exterior walls.

As for ceilings, the same is true—explore using a material with a higher R value. Again, traditional methods call for drywall or plaster ceilings in most homes. The real increase in thermal resistance for roofs and ceilings is in the fiberglass insulation that is placed in the attic above. However, since heat rises and the greater the R value, the better for the ceiling and roof, it is worth looking into substitute materials for the ceiling. Wood fiberboards, corkboards, and other fiberboards can provide a much higher R value at the ceiling and yield some positive acoustical effects as a by-product of their use. Similarly, the use of sheathing, a wood fiber material, on exterior walls is good not only for the extra insulating value but also for the rigidity that it gives the building's frame. Many home builders do not use sheathing but only use a building wrap. This keeps their costs down, so the home owner should always ask how the exterior walls are built. After all, the home as a shelter means that the exterior walls and the superstructure must resist the elements, and the home owner should know what materials are protecting his or her family.

Building Wrap to Reduce Infiltration

Building wrap is used for many reasons, but basically it is another level of protection from the outdoors. By increasing the insulation in the exterior walls and ceilings, adding storm windows and doors, and using high-efficiency heating/cooling equipment to make the house energy efficient, it is still necessary to include a building wrap. Building wrap is not a gift-wrapping of the residence on completion but rather is a critical element in the middle stages of the project. If it is left out, the home owner could see high energy bills in the future due to increased air infiltration. An unwrapped house can undermine the comforts and savings for which the home owner is paying.

Air penetration can cause draftiness and inside temperature fluctuations. If moisture vapors cannot get out, the increased humidity may cause great discomfort and damage to construction materials within the wall cavity. Building wrap provides a degree of air penetration resistance, water penetration resistance, and moisture vapor transmission. As can be seen in Fig. 7.1, the building wrap is installed after the foundation and frame are in place. Many builders will place the building wrap over the studs with no

■ **7.1** *Building wrap to reduce air infiltration.* (Courtesy of Wayne J. Henchar Custom Homes.)

particleboard or sheathing. These builders believe the siding and interior finish boards will provide the structural rigidity when in place. However, building wrap by itself provides no structural support and may be exposed for months until the siding goes on, often a final-stage process. An ideal building wrap should be easy to install and be translucent if not almost transparent so that one can actually see through it. This will eliminate the guesswork on cutting, fitting, and fastening. The sheet should reduce glare when installed and give protection from damage by ultraviolet rays and winds. Once covered up by the exterior finish products, the building wrap will pay for itself in energy savings probably within the first year.

Besides full building wraps, vapor barriers of typically 4 mil thickness should be considered. They are a cross-laminated sheeting made from high-density resins. They are placed between the insulation and the interior wall board. Below grade, the vapor barrier should be installed around the outside perimeter from the base to the ground level and laid under the concrete flooring to eliminate moisture migration.

Air Locks

A great deal of outside air comes into a home when occupants enter and exit the residence. This is why an air lock of some sort is well worth the effort in the design and build stages of the project. Not only will the air lock keep out the cold winter air, but it also will keep out the extra moisture-laiden air of summer. Figure 7.2 shows a couple of air lock schemes. Basically, an *air lock* is any physical break in the full, direct path of the air

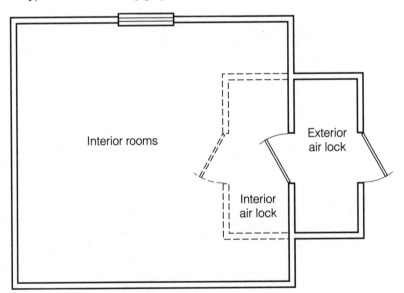

■ **7.2** *Inside and outside air locks.*

entering or leaving a house. One door is closed, in theory, after the person enters, and then the second door is opened. This minimizes an in-rush of cold air. Depending on how the house is pressurized, this in-rush of air can be substantial. Air locks sometimes do not work because both doors are left open simultaneously and the air has a direct path into the house anyway. As a rule, however, the air lock is used often and works very well to save on home heating and cooling costs. One type, the revolving door shown in Fig. 7.3, creates an air lock and is used in many commercial and apartment buildings. Revolving doors are not practical from a cost standpoint in a home. We all know enough not to leave doors to the outside open when the weather is very humid, but keeping doors to the outside closed when not in use will not keep all outside moisture from coming in. Additionally, moisture is created within the home by cooking, bathing, and human exhalation.

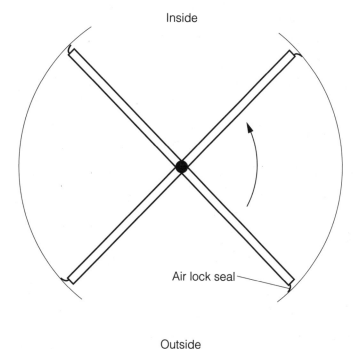

■ **7.3** *Revolving doors.*

Even a tightly built, weather-stripped house with an air lock can have one-half to one full air change per hour. That is, new air can come in through tiny cracks and crevices around windows and doors and through utility outlets and the fireplace. Exhaust fans in bathrooms and the kitchen can step up the air change process. Older homes sometimes have as many as two air changes each hour. If the outside air is cool, it will be drier than inside air. On the other hand, during warm, humid weather, each air change can bring in 4 to 5 pints of moisture. During a day, 10 to 20 gallons of moisture can be brought into a home through infiltration. This excess moisture contributes to mildew problems and requires energy to remove. By keeping the humidity levels controlled within the house, the cooling system does not have to run as much. Humidity levels in check also reduce the need for dehumidifiers to lower water vapor levels.

Except in home upgrade and retrofit applications, air locks do not have to be an afterthought. They can be incorporated in the design stage of the residence. Ironically, the larger and grander the house, the less the air lock is employed. The entryways into these houses are virtually the home owner's statement about the house, and an air lock is seen to detract from that beauty. Unfortunately, these larger houses have high ceilings, large doors, and elaborate windows that make for tremendous heat loss. On the other hand, these home owners usually are able to afford the higher energy bills (even though this methodology is defeating the purpose of an energy-efficiency book and trying to save energy within the home overall).

Reducing Air Infiltration with Tapes and Sealants

The most important step in air infiltration control is sealing cracks and holes, and the first place to check is around the windows and doors. These leaks can be sealed using caulking or expanding spray foam sealant. For cracks that are ¼ in or less, use siliconized caulk. Use spray foam sealant for larger cracks and holes. In shim spaces around windows and doors, use low expanding foam. In addition to leaks around windows and doors, check for poorly installed and leaky ductwork in hidden passages in your walls, called *air passages*. These are major contributors to drafts and uncomfortable room temperatures. Sealing and insulating ductwork and increasing the levels of insulation in your attic can greatly enhance the efficiency of your cooling and heating system as well as make your home more comfortable. Air leakage in the home can be found around electrical outlets, the plumbing, many duct work joints, windows, and doors.

On the interior of a building, tape is recommended to make the vapor barrier airtight. The vapor barrier should be overlapped by 6 in and taped along the overlaps. The sill plate, header joints, cracks around windows and doors and around electrical outlets, as well as piping and ductwork that penetrates the vapor barrier should all be taped.

"Green" Products

A large number and variety of products can be used in a home to help save energy, and some also come from recycled sources. These products come from energy-efficient environments. When one stops to consider that the average wood-framed house uses an astonishing 11,000 board feet of lumber, it is no wonder that some of these resources are going to become scarce in the future. Siding can be made from recovered wood fibers and cement. The exterior trim for houses can be made from wood composites using recovered wood fiber. Window frames can be made of energy-efficient fiberglass and insulated with fiberglass. As a matter of fact, many conventional framing materials could be made from plastics and fiberglass products. The thermal resistance values of these products is high, and the overall effect on the environment would be positive. Recycled automobile tires can be used as carpet padding; New types of concrete can be made from recycled coal combustion fly ash, similar to the drywall boards being manufactured today. As long as this hazardous material is embedded and fused inside concrete or a pressed board, it is safe to use. Floor tiles can be manufactured from recycled automobile windshield glass or from old, discarded fluorescent light bulbs. There are hundreds of alternatives to the traditional building methods and materials used in the home construction industry. The home owner can request that recycled and "green" products be used wherever possible in his or her residence. Some other products available are as follows:

The Chromalux full-spectrum bulb provides increased color rendition, described as the closest thing to natural light. These bulbs are designed to last up to twice as long as ordinary incandescent bulbs. Chromalux bulbs are made of special hand-blown glass that is not colored or coated but instead contains neodymium, a rare earth element used in lasers. This causes the light to be purified, allowing the violet, blue, green, and red components to be strengthened without producing an imbalance of one color over another. A 60-W bulb can cost about $10. There is a trend toward high-efficiency, long-lasting light bulbs. In the future this may be all that one will be able to purchase.

Inner rooms of a house that have no exterior walls cannot have a window. Basements could use a little more light, but there has to be means of getting the natural light from the sun to the area. A product called a *sun pipe* can brighten any room in a house with natural daylight at a fraction of the cost of a standard skylight. The Sun Pipe reflects natural daylight from above the roof, through the attic, and directly to a light-diffusing dome on the ceiling of the room needing the light. The heat also stays outside. Illumination on sunny days is equivalent to 1500+ W of incandescent light, 200 to 900+ W on cloudy days! It works much like fiberoptics but on a larger scale. Properly installed, a Sun Pipe should never leak. They typically come with standard flashing and a storm collar. The aluminum Sun Pipe can be purchased in various diameters and should require no cutting of joists or rafters and no framing, taping, drywalling, or finishing. Installation is relatively simple, depending on the sunlight's final destination. Options include a 2 by 4 ft extension that allows sections to be piped together and a device called a Sun Scoop that can capture up to 170 percent more sun during the winter solstice. Kits include a 4-ft pipe, clear-top dome, white-bottom dome, sealing materials, installation hardware, flashing collar assembly, and installation guide.

Whenever selecting materials for a house, the home owner should consider recycled or "green" products. Often we are not even aware of their existence, since they emerge quickly and are not always advertised well. The home builder should educate himself or herself on the latest recycling trends and products. Recycling of glass, plastics, and paper is gaining more momentum each year, and new products are readily becoming available. Most of the time we cannot even tell that the composition of a particular material has a percentage of recycled components within it. Therefore, why not attempt to use them wherever possible in our homes and do our part to help the environment along the way? After all, energy efficiency is ultimately using less energy to allow the creation of less in order to save the environment. It takes energy to create building materials; moreover, it takes energy to collect the raw materials to manufacture the products. If we reuse some of the products that are being discarded, we are eliminating the step of collecting the raw materials. This alone saves energy.

Finishing Remarks

We all want our homes to look great. Our home is a statement about ourselves and our successes in life. Our home also can look good and save energy at the same time—if this concept is designed in. We can no longer talk about energy savings and live lifestyles of wasting that very energy. Federal regulations and laws are only going to get more severe as time goes on with respect to energy efficiency in residential construction. Therefore, the time is now to design all the way to the "finish" with energy-efficient methods and materials.

So many times our houses are designed with proper orientation to the sun, increased R values of insulation, and high-efficiency appliances, but the final stages of construction—the finishing systems—are overlooked. Make it look good or make it look acceptable so that we can leave the job site is the final thought on the minds of all—the home builder, contractors, and even the home owner. Building projects get delayed or just seem to last forever. Everyone wants to move on to the next project, and thus "let's just get this one done!" More upfront planning has to go into the finishing of the house and to the colors of materials and actual types of materials used in the final stages of the house's construction.

The finishing of the house is often left up to the home owner. The home builder usually finishes the walls and slaps a coat of white paint on and that is it, unless the contract has provisions for wallpaper or other special finishes. Even then, the thought that goes into the finish system and the colors is typically not around any extra energy savings or possible heat gain. This may happen only because the home owner and home builder are not thinking along these lines, and they could. If the theme of the house is energy efficiency, which is more becoming a federal mandate, then the design process from beginning to end and from front to back should strive for every bit of energy savings possible every step of the way.

Energy savings by virtue of the finishing systems has to be discussed in the early design stages so that when it comes time to finish the house, materials are available and provisions have been made for these special systems. Ironically, the incorporation of energy-efficient methods into the finishing process is not a cost premium situation. It does not cost any more for a darker color paint than it does for a lighter color. Locating a window within a room in a certain spot to let most of the daylight in does not cost any more if it is planned for ahead of time. Even substituting a higher-thermal-resistance material may cost a little more, but if the plan is in place to use something along these lines, then one can shop around for the best material deal while the house is being excavated, framed, and built—there will be an ample amount of time.

During the design stages, make a list; better yet, use a finishing system checklist similar to the one shown in Fig. 7.4. For every room in the house determine how the windows should be oriented with respect to the sun, and locate them accordingly. What type of surface will be employed on the exterior walls and the interior walls that see the sunlight? Will shades, shutters, and blinds be used, and where? Even if the application is a home retrofit, these checklist items need to be addressed. All that is required is some dialogue and discussion along the lines of an energy-efficient finishing system. Home owners and home builders have got to open up discussions along these lines or else they are overlooked, and usually it is too late and too costly to incorporate them later during the actual construction. Just ask the questions!

The home builder and contractors are being asked today to be more than builders and contractors. They have to be more creative with building designs, more knowledgeable about energy savings, and more up-to-date on new products. No longer can they perform the basic functions of their trade as in years past. The building construction world has changed, and so must they. The successful home builder will adapt to these expectations in order to keep his or her clients, the home owners, happy. All disciplines in the twenty-first century will be expected to contribute more than was expected in the past. Every aspect of the building process has to be scrutinized to achieve the optimal in aesthetics and energy efficiency. Finishing systems and new energy-efficient materials can be an area worth exploring in realizing the efficiencies and paybacks home owners and the government are looking for.

Finishing systems checklist:				
Room	Orientation	Wall materials	Wall colors	Windows/door count
Living room				
Kitchen				
Dining room				
Family room				
Bedroom one				
Bedroom two				
Bedroom three				
Sunroom				
Other				
Other				

■ **7.4** *Finishing systems checklist.*

Residential Mechanical Systems

8

Within the typical residence there are several components that can be classified as mechanical in operation. Either there is a fluid being moved, such as a liquid or air, a device causing something to compress or change position, something in motion, or any number of other types of systems that are not 100 percent electrical in operation. Energy savings can be attributed to many of these devices that are not calculated directly in a payback analysis. Any of the "rotating or moving objects" use some type of energy. Most of the time this energy is electrical. Hence it is necessary to convert from electrical to mechanical. More important, it is necessary to understand what makes these mechanical systems "tick" and how efficient they are. This understanding will better allow the home owner and home builder to select the most energy-efficient mechanical system.

Types of Mechanical Systems

Frequently, these mechanical systems are referred to as the heating, ventilating, and air-conditioning (HVAC) systems. This acronym encompasses a wide variety of equipment and disciplines. While the term *mechanical* might mean something different to some home owners, the same can be said of the term *HVAC*. Some home owners may consider an HVAC system to be a machine with noisy, clanking parts. Others will view anything mechanical as anything nonelectrical. All are right, and the reality is that there are varying degrees to these definitions. The typical home's mechanical systems are grouped in this manner: heating and cooling systems, water heating, plumbing, ductwork systems, and humidifying and dehumidifying systems. Mechanical systems also can be classified as dynamic and static. A *dynamic* system is one in which something is moving or changing. A *static* system is one in which there is no apparent movement and change is gradual. Condensation and evaporation of water vapor in a home are actually a mechanical system at work! Alternative energy–sourced equipment, other than by electricity, would include equipment powered by natural gas, oil, or solar energy, and this equipment also could be grouped with the mechanical systems of the residence.

Nevertheless, the mechanical systems within a residence by and large are powered by electricity. Even the heating or cooling system, whose primary energy source is nonelectrical, still needs an electrically powered fan or pump motor to deliver the heated medium. These systems could be labeled with the more modern term for electrical and mechanical systems, namely, *electromechanical*. Whatever the terminology, the bottom line is that these systems must perform reliably and energy efficiently.

HVAC systems also can be classified by the medium by which heating or cooling is accomplished. An air-to-water HVAC system is a type of central HVAC system that distributes the conditioning effect by means of heated or chilled water and heated or cooled air. In an all-air HVAC system, distribution of the conditioning effect is solely by means of heated or cooled air. An all-water HVAC system is a type of central HVAC system that distributes the conditioning effect solely by means of heated or chilled water. HVAC systems are designed around zones. *Zones* are areas of a building that must be controlled separately if conditions conducive to thermal comfort are to be provided by an HVAC system. Often, a zone is an area that is controlled from a single control point, usually a thermostat. In a variable-air-volume (VAV) HVAC system, the central air-conditioning system uses a single supply air stream and a terminal device at each zone to provide appropriate thermal conditions through control of the quantity of air supplied to the zone. A local HVAC system produces a heating or cooling effect in or adjacent to a space that requires conditioning. Distribution of the heating or cooling effect is effectively limited to a single space.

The most common approach to the heating and cooling of a residence is a central HVAC system. This is a system that produces a heating or cooling effect in a central location for subsequent distribution to satellite spaces that require conditioning. This type of system incorporates all the ductwork, the piping for any supply and waste fluids, and the electrical controls and wiring. In a multizone HVAC system, a central air system uses an individual supply air stream for each zone. Warm and cool air are mixed at the air-handling unit to provide supply air appropriate to the needs of each zone. A multizone system also requires the use of several separate supply air ducts.

Efficiency

In any mechanical system, some energy is used as work, and some is lost. As illustrated in Fig. 8.1, the power entering the system is not all used for work. Work is done, and this is what we would perceive as the amount of

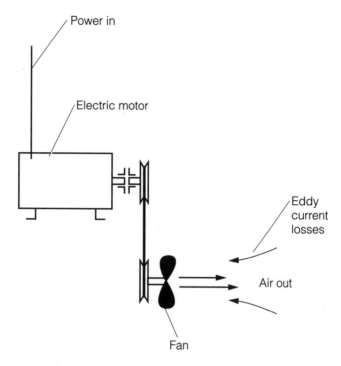

■ **8.1** *Work and energy in a common household
fan system.*

energy or power used, right? Not so, since the inefficiency of the system
used some extra power while getting the work done. The problem of poor
efficiency or inefficiency is difficult to solve. One reason is that there are
usually several components to any mechanical system, each with its own
contribution, or lack thereof, to the overall efficiency of the system. There
are friction, noise, and heat losses. Equipment also has optimal conditions
for its best performance and therefore its best efficiency; however, the
equipment may not always be run under these circumstances. Efficiency is
calculated fairly easily: Divide the output power of a system by the input
power (Fig. 8.2). The output power value always should be lower than the
input value. There is not any system, mechanical or electrical, that is 100
percent efficient. Some losses always occur.

Systematic studies have been done on the relationships between energy,
heat, temperature, and work. These studies have given us the laws of thermodynamics, and these laws govern much of the operation of our home
heating and cooling systems. An important concept in thermodynamics is
entropy. Entropy—not to be confused with *enthalpy*, a measure of the relative heat content of air measured in British thermal units per pound of dry
air—allows us to measure usable and nonusable energy. Controlling energy
so as to do useful work can be likened to a 12-V battery, which stores its en-

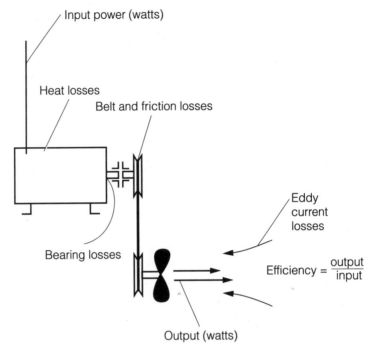

Input power (watts)

Heat losses

Belt and friction losses

Eddy current losses

Bearing losses

$$\text{Efficiency} = \frac{\text{output}}{\text{input}}$$

Output (watts)

■ **8.2** *The efficiency of a system with associated losses.*

ergy, as opposed to a fireplace, in which heat energy is lost to the surrounding atmosphere (or up the chimney). Manufacturers of heating equipment are continually striving to produce the most efficient package available and to get the best control of the entropy of the system.

The losses within a given system are the inefficiencies. They usually occur in the form of lost heat energy, but they can be in the form of retarding motion (friction), audible noise energy, and unwanted wear and tear on the system's components. Thus, whenever a particular system is being evaluated for efficiency, do not rely just on the efficiency rating on the label. Frequently the efficiency percentage is misleading because the testing was done in a very controlled environment and under optimal conditions. Typically this is not what your home is going to provide. Use efficiency data as a guide, not as the full, determining factor when selecting equipment. Ultimately, the initial and operating costs will have to be weighed against the performance, and this is usually never fully monitored by the home owner. The EnergyGuide labels and other standards provide additional guidance for selection. Finally, the builder should provide what he or she feels is the best equipment package to the client.

The efficiency of the heat source must be taken into account. Gas furnaces range from 70 to 80 percent efficient, measured seasonally over an entire year of operation. Gas-fired furnace and boiler units with efficiencies of 90 to 95 percent are also available but usually are produced in large output sizes and thus are more expensive and may not be applicable. Two important acronyms used relative to efficiency are *SEER* and *AFUE*. AFUE is the *annual fuel utilization efficiency* and SEER is the *seasonal energy efficiency ratio*. Both are discussed in relation to pertinent equipment later on in this chapter.

Cooling Systems

Cooling is a process that removes sensible or latent heat from a designated space. Typically, residential systems that are capable of delivering both heating and cooling are sized around the house's cooling loads. The *cooling load* is the magnitude of heat removal required to maintain a building at appropriate thermal conditions. Homes today are built with the intent to absorb as much energy from the sun during the day as possible. These houses are constructed with sufficient insulation to retain most of that heat gained, while several other measures are taken to ensure that heat is not lost wastefully. This is especially true in climates that have a heating season. Therefore, whenever selecting a cooling system, the energy required to cool a house, designed to retain heat, is usually greater than that required to heat it in extremely cold and harsh conditions. Most of today's cooling systems operate using a compressed refrigerant to move heat from a given area. This is called the *compressor cycle* and is reviewed later in this chapter. Room air conditioners use the standard compressor cycle and are sized to cool just one room of a house. To cool an entire house, several individual room units are necessary. Using several room units is not practical, and therefore, a central air-conditioning system sometimes is employed. Central air-conditioning systems also operate on the compressor-cycle principle and are designed to cool an entire house. The cooled air is distributed throughout the house using air ductwork, which is commonly the same ductwork used by the heating system. Typically, the sizing of any cooling system is in units called *tons*, and the capacity of all cooling equipment is measured in British thermal units of heat-removal capacity per hour. A ton of refrigeration is the amount of cooling provided by melting a ton of ice in 24 h. This is a cooling rate of 12,000 Btuh.

Another type of cooling system that does not use the compressor cycle is called an *evaporative cooler*. Sometimes also called *swamp coolers*, these systems cool the air by blowing it over a wet surface. We have all experienced this phenomenon when we get out of a swimming pool while a breeze is blowing. As the water evaporates, it absorbs heat from the air. A big factor in the practical use of evaporative cooling systems is the ability of the air to absorb moisture. Dry air, found mainly in the southwestern states such as Texas, New Mexico, and Arizona, is the perfect medium for these evaporative coolers.

Often the cooling system can be set up such that an economizer is implemented. An *economizer* provides a means of space cooling through the introduction of outside air rather than through refrigeration. In so doing, energy is not used to provide the cooling, but rather, cooler outside air is introduced to the system. An economizer cycle allows outside air to be used for cooling when exterior conditions are appropriate. Keeping track of outside air temperatures and humidity allows for economizer circuits to be employed. Home automation systems can perform this function well.

Once the cooling loads have been determined and the choice of cooling system has been made, either using a compressor cycle or an evaporative type, then a further analysis of the different mechanical cooling units is necessary. Heat pumps and geothermal systems are capable of heating and cooling a structure as well as other mechanical systems. Each is reviewed in some detail later in this chapter.

Air Conditioners

The term *air conditioner* for years usually has implied that the system is for cooling purposes only. However, the term *air conditioner* means just a bit more than this; it really means a device used to condition the air either by heating it or by cooling it to some desired temperature or a device that simultaneously controls the temperature, moisture content, distribution, and quality of air. Most air conditioners work on the principle of the compressor cycle, the same as a refrigerator uses. A compressor is a device designed to increase the density of a compressible fluid, and in a refrigerator or air conditioner this is the component that compresses the refrigerant.

A house can be compared with a refrigerator box, as is seen in Fig. 8.3, and the heat is removed from within the "box" out to a heat sink. The air-conditioning system uses a fluid called a *refrigerant* to transfer heat from indoors to outdoors. The refrigerant is compressed outdoors, where it releases heat, and evaporated indoors, where it absorbs heat. The process of moving heat in this manner requires input energy in the form of electricity to drive it. Electrical power is used both to pump the fluid through the cycle and to circulate the cooled air around the home.

Even with the most efficient air conditioners, it makes a great deal of sense to do everything you can to reduce air-conditioning loads. Just a few conservation measures are often so effective that houses in the northern half of the country and in mountainous regions can get by without air-conditioning on all but the very hottest days. If you are planning to buy a new system, reducing the cooling load will save you a lot of money right away by letting you buy a smaller, less expensive system. People are comfortable during the summer at temperatures between 72 and 78°F with a relative humidity of 30 to 60 percent. If the relative humidity climbs, then the com-

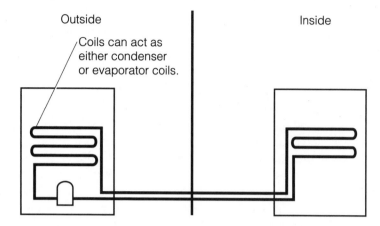

Outside   Inside

Coils can act as
either condenser
or evaporator coils.

Flow direction of the refrigerant reverses in seasonal changes.
Heat sink is outside in summer and inside during winter.

■ **8.3** *A house with the outside environment as the heat sink.*

fortable temperature range has to be lower. The comfort range can be increased to 82°F with a small amount of air movement, such as that provided by ceiling fans. Increasing the comfort range means you can maintain comfort while running your air-conditioning system less and saving money.

Heating Systems

Heating systems have come a long way from the huge, crackling bonfires that heated the city of Ushuaia, Tiera del Fuego (which means the "land of fire") in Argentina over 100 years ago. Granted, Tiera del Fuego is 54 degrees south of the Equator (a comparable latitude to Alaska), and technology will tend to take longer to get there, but we all started with that basic heat source—fire. Today we have developed very efficient ways to generate, use, conserve, and recoup heat. Of all the aspects of building construction, heating systems seem to get a great deal of attention. Let's face it, we don't like to be cold! The next decade promises to produce even better products for heating.

In new home construction, several options are available to the home owner and home builder when it comes to a heating system. The limitations of having existing plumbing, ductwork, or wiring in place, as in the case of an existing home, do not present themselves. Rather, a home that is partly laid out around an optimal heating system is practical with new construction. Where to place the primary heat generator, how to route plumbing and ductwork, and sizing become much simpler when thought out on paper and thoroughly discussed between home owner and home builder.

When it comes to heating systems, there are many options. The fuel source has first got to be established. Analyze the costs in your area for certain types of fuel and its availability. If you currently have electric resistance heat, switching to oil or gas heating systems usually will save you money. If you have an oil or gas system, you probably want to continue to use the same type of fuel because the storage tank and pipes are already in place, but sometimes it makes sense to switch.

Before the energy crunch of the late 1970s, most heating systems were oversized, meaning their capacity to produce heat exceeded the maximum projected heating requirements of the house. Some systems were oversized by a factor of 2 or 3. It is important to note that oversized systems do not operate at peak efficiency and can cost more to purchase, operate, and maintain. To be most efficient, your heating system should be able to keep your home at 70°F on the coldest day of the year. A heat-loss analysis can help you determine what size heating system you will need to be able to do this. The analysis should tell you what your peak hourly heating demand will be throughout the year, measured in British thermal units per hour. The British thermal unit is a measure of thermal energy (heat) and is equal to the quantity of energy required to raise the temperature of 1 lb of water by 1°F. Typically, a new heating system should not exceed the peak hourly heating demand by more than 25 percent.

Today, energy-efficient homes have very little heat loss because of high insulation levels and airtight construction. This leads to two problems for the heating system. It is harder to properly size a heat source and then to provide adequate ventilation to maintain indoor air quality. With builders focusing on the structure's orientation and energy-efficient construction, a residence can practically heat itself. Except for those extremely cold periods, a properly designed and constructed energy-efficient home sometimes can gain almost enough daily heat from waste sources such as the heat given off by lights, people, and appliances. During sunny, cold days, solar energy gains also contribute to reducing the heating load. These heat sources often are called *internal gains*. Modern control systems such as programmable thermostats and complete home automation systems can further help to reduce heating energy consumption.

The heating equipment in a residence must be capable of maintaining an interior temperature of 68 to 72°F (20 to 22°C) during the heating season. This leads to frequent on and off operation, thus reducing both efficiency of fuel use and service life of heating equipment. If a fuel-burning furnace, boiler, and/or hot water heater is required, building an airtight enclosure (mechanical room) around the appliances can help control chimney heat loss. Any fuel-burning process needs air and oxygen to perform combustion. These systems will pull the air from virtually anywhere, and the net result is that valuable heated air is usually exhausted through the flue with the dangerous carbon monoxide and other gases. Separate fresh-air supplies should be fed into this room or to the furnace itself. No previously heated air is used by the fuel-burning system, and cold outside air is prevented from entering other areas of the home. This isolated room must be insulated and sealed from the rest of the house. Water pipes and heat supply ducts also should be insulated.

Often, if the heating and/or cooling system are noncombustible types, the mechanical equipment still is grouped together. As can be seen in Fig. 8.4, locating hot water tanks with the heating and cooling equipment is almost always done. This approach usually minimizes lengthy runs of piping and electrical wiring. In addition, if changes in equipment have to be made in the future, most of the equipment is readily accessible.

■ **8.4** *Electric hot water tank located next to a heat pump.*
(Courtesy of Wayne J. Henchar Custom Homes.)

Forced-Air Systems

Forced-air heating systems are very efficient at distributing heat around a home, preventing stagnation of air, and moving heat from different sources to the overall space. They also work well with mechanical ventilation systems. The central heat source could be a fuel-fired furnace. Unfortunately, the smallest sizes usually available are in the 50,000 to 60,000 Btuh range, and this heating capacity is often much too large for an energy-efficient home. Using such a large heat source is inefficient in terms of fuel consumption. For example, a 50 percent oversized furnace will use 20 percent more fuel in heating the same space than a correctly sized unit. Even heat distribution, air movement, filtering capability, humidity control, and low maintenance are some of the advantages of a properly designed and installed forced-air heating system.

A forced-air distribution system works well if a home receives additional energy from the sun. Since passively heated spaces can easily overheat when the window area is too large or if there is not enough mass to absorb and store the solar energy, having continuously circulating air with the forced-air distribution fan running at a slow speed helps prevent overheating. This a good application of a variable-frequency ac drive and is discussed further in Chap. 9. Passively heated air is distributed to all the spaces in the home, not just to those on the south side.

Gas Furnaces

As far as fossil fuels go, natural gas is probably the best option available to the home owner for heating purposes. Natural gas is usually available, either by direct pipeline into the home from the gas company or by the installation of a natural gas storage vessel on the property. In either case, the cost of the natural gas makes it attractive as a heat source, as do its efficient, clean-burning properties. The issues to address are leak detection and prevention, the exhausting of combustion-related by-products, and the need to add some type of cooling system.

High-efficiency, forced-air gas furnaces offer efficiencies of 90 percent or better. These units use electronic ignition, induced-draft fans, and condensing heat exchangers. Ductwork and installation are similar to a standard furnace with the exception of the chimney and condensate drain. Condensing furnaces require a drain pipe connected to a floor drain to allow condensation (water) from the heat exchanger to drain. A standard chimney is not required because the exhaust air temperature is reduced such that high-temperature plastic pipe can be used as an exhaust vent out a sidewall. A fresh-air duct from the outside can be used, ducted directly into the cold air return. Combustion air is separately ducted from the exterior to the front of the furnace.

Hot Water Heating Systems

The name given to the science of heating or cooling with water is *hydronics*. Many early hydronic heating systems operated on the principle of gravity. Water expands when it is heated, and this makes a given volume of heated water lighter in weight, and thus it rises above the cooler water around it. Using the principles of convection and radiation as the heat-transfer method, water can be placed in circulation inside a contained and heated system. As can be seen in Fig. 8.5, water gets heated in a boiler, and this heated water rises through the appropriate riser piping. The heat-transfer method to this point is convection. The water rises to a radiator and gives off heat by radiation to the space around the radiator until the water's actual temperature drops, usually from 200 to 180°F. The water's volume shrinks as it cools, and the water becomes much heavier. At this point, this heavier water falls back to the boiler via the return piping as a result of gravity, and the cycle is repeated.

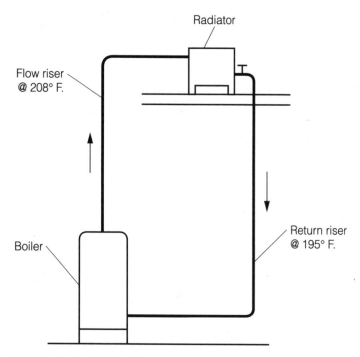

■ **8.5** *Hot water heating.*

Many older homes still have such systems in place and operating; however, technology has allowed us to positively modify these hot water systems. A circulating pump has been added to the system, thus making it a forced-circulation hot water heating system, as shown in Fig. 8.6. This pump, or circulator, allows for more radiators to be added and gives the overall sys-

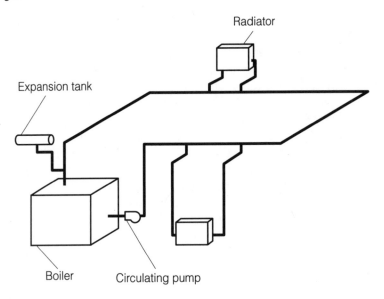

Radiator

Expansion tank

Boiler   Circulating pump

■ **8.6** *Hot water heating with a circulating pump.*

tem better performance. Additionally, with a pump system, radiators can be placed such that gravity does not have to be relied on for water return. With temperature and pressure sensors and even motor controls, a hot water heating system such as this can be controlled exactly by any home automation system. This hot water system is actually a good candidate for a home automation system because, notoriously, the hot water needs time to be heated, be delivered to the radiators, and actually heat the surrounding air—typically taking longer than a forced-air system, which can deliver heated air almost instantly. A radiator approach, by design, delivers heat a little more slowly. However, a home automation system can monitor the house and temperatures, inside and outside, to predict when to turn the boiler on. This can start the process ahead of the demand.

Auxiliary Units

Frequently, climate and temperature demands are such that the entire heating system does not have to be working, but rather, some form of localized, ambient heat is preferred, or else a smaller heating device is integrated into the existing heating system. Using the advantages of a forced-air system, you can add a small auxiliary heat source. This could be an electric heating element (not as cost-effective, but evaluate the total hours of operation), a hot water heated-coil unit (boiler has to work), a heat pump (convenient and cost-effective), or a separate wood-burning system. Separate, electrically powered fan coil units also can be used and sometimes are used in quantity in office buildings where piping is available for heated or chilled water. The fan coil is a small-scale air-handling unit, including a fan, a coil or coils, and a filter, that is normally located in or directly adjacent to a conditioned space. It also serves as a delivery device in an all-water or air-water central HVAC system.

With an auxiliary heating scheme, heat is picked up and distributes via the forced-air system. Hot water heating can be combined with a forced-air system. Hot water from a boiler is circulated through a radiator placed in the ducting of the forced-air system. A fan forces the air through the radiator, where the heat is picked up and distributed throughout the entire system. The ductwork remains the same. An air-to-air heat exchanger maintains indoor air quality by furnishing fresh air, which is preheated, to the cold air return. This is then distributed to the rest of the house by the forced-air system.

Refrigeration and Freezer Systems

With many homes having one and sometimes two refrigerators, as well as typically an additional freezer, these appliances constitute a mechanical system to the home owner. Understanding the principles of refrigeration will better allow the home owner to understand the work being done and the energy being consumed.

The basic components of a typical refrigeration system are also embedded into many other household systems such as heat pumps and geothermal systems. Besides the refrigerator enclosure, the mechanical components, and the electric circuits, the system works around the functions of a few other subsystems. As can be seen in Fig. 8.7, the subsystems all have their own jobs to do. The enclosure supports the evaporator and condensing unit. In the evaporator, a liquid refrigerant expands and becomes a vapor. The refrigerant vapor, in this expanded state, absorbs heat from the foods within the refrigerator enclosure. It transfers this heat out of the internal part of the enclosure at the condenser and is compressed again to repeat the cycle. The heat-pump system operates basically the same way with the same basic components: a compressor, a condenser, refrigerant, and an evaporator. The difference with the heat pump is that it can reverse the cycle and cool or heat, whichever it is directed to do.

Today's refrigerator and freezer systems have evolved to much more efficient and maintenance-free devices. Although the basic concepts for heat transfer are still applicable, the means of getting the work done have changed. Environmental issues have forced a change in the types of refrigerants allowed to be used in the compression cycle. Because the refrigerant is the heat-transfer fluid employed by the refrigerating process, it is selected for its chemical properties, which include stability, high thermal

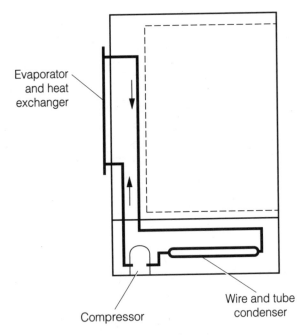

Evaporator
and heat
exchanger

Wire and tube
condenser

Compressor

■ **8.7** *Refrigeration subsystem.*

capacity, low viscosity, and high/low-temperature change points. Freon was
the main refrigerant for years. However, with growing concerns about what
hydrofluorocarbons (HFCs) and chlorofluorocarbons (CFCs) are doing to
the ozone layer, many substitutes are appearing. A list of various types of
refrigerants and their scientific names is given in Table 8.1.

■ **Table 8.1 Various Refrigerants
and Their Scientific Names**

Refrigerant Abbreviation	Refrigerant Scientific Name
R-11	Trichlorofluoromethane
R-12	Dichlorodifluoromethane
R-13	Chlorotrifluoromethane
R-14	Tetrafluoromethane
R-22	Chlorodifluoromethane
R-23	Trifluoromethane
R-113	Trichlorotrifluoroethane
R-114	Dichlorotetrafluoroethane
R-116	Hexafluoroethane
R-123	Dichlorotrifluoroethane
R-124	Chlorotetrafluoroethane
R-134a	Tetrafluoroethane

Freon comprises a group of compounds containing carbon and fluorine with other elements such as chlorine and hydrogen—hence the names *chlorofluorocarbon* (CFC) and *hydrofluorocarbon* (HFC). They are non-flammable, odorless, and colorless and are low in toxins, enough that they have been used in refrigeration and cooling systems for years. Freon has a low boiling point, which makes it well suited for use as a closed-loop refrigerant, able to be compressed and evaporated in a heating/cooling cycle. The most common grades of Freon used in the past were R-11, or Freon 11, R-12, or Freon 12, and R-22, Freon 22. At present, there is a worldwide ban on their use, and thus many substitutes are being tried, some more toxic than others to humans. It is advisable to ask about the type of refrigerant being used in any cooling or refrigeration system going into your home. Also ask about leak-detection methods concerning that appliance.

As refrigeration systems continue to evolve both as environmentally friendly and as high-tech, energy-efficient packages, their use continues to increase. Many homes have two refrigerators (sometimes more) and at least one freezer. This equipment does have an impact on the electrical energy load of the residence and should be monitored. Refrigeration technology has brought us many convenient innovations. Drip and water pans are virtually a thing of the past. No more defrosting, better and separate compartment temperature control, and a heat-exchange system that does not require bulky external evaporator coils are just a few of the improvements. Today, ice makers and cold water dispensers and other amenities are making their way into refrigerator design. How far we have come from the block-of-ice systems of yesterday!

Heat-Pump Systems

These systems can go both ways—heating or cooling—and this is what makes them so attractive to the home builder and home owner. This ability to act as a true air conditioner, winter and summer, also makes this device a very efficient system. Additionally, since it does not require fossil fuels because it does not work on the principle of combustion, it is attractive environmentally. As is shown in Fig. 8.8, a heat-pump system can reverse its operation depending on the season or preference of the home owner. By reversing the cycle of operation, the heat sink is either the outdoors or the indoors. Heat pumps are very energy efficient. The heat-pump technology of today has persevered over several years. Even at temperatures we consider to be cold, air, ground, and water all contain useful heat that is continuously replenished by the sun. By applying a little more energy to the process, a heat pump can add a little energy to this heat energy to bring it to the level needed to deliver comfortable heat. Industrially, heat pumps

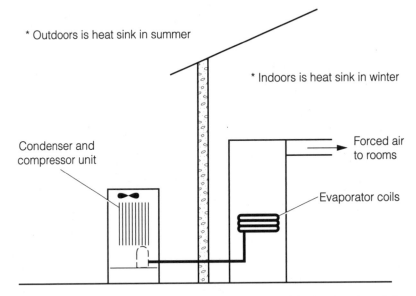

* Outdoors is heat sink in summer

* Indoors is heat sink in winter

Condenser and
compressor unit

Forced air
to rooms

Evaporator coils

■ **8.8** *Heat pump reversing to make the heat sink the interior of the house.*

also can use waste heat sources, such as from industrial processes, cooling equipment, or ventilation air extracted from buildings. A typical electric heat pump will need just 50 kWh of power to turn 100 kWh of freely available environmental or waste heat into 150 kWh of useful heat.

Through this unique ability, heat pumps can dramatically improve the energy efficiency and environmental value of any heating system that is driven by primary energy resources such as fuel or power. Considering the fact that heat pumps can meet space heating, hot water heating, and cooling needs in all types of buildings, as well as many industrial heating requirements, then heat pumps are a clear-cut choice, as residential heating/cooling devices. In some regions of the world, heat pumps already play an important role in energy systems. However, if this technology is to achieve more widespread use, a decisive effort is needed to stimulate heat-pump markets and to further develop the technology. Heat-pump technology has remained intact but has not seen any new developments over the past few years. However, it is important to note that many local governments and utilities strongly support and encourage the use of heat pumps. In all cases it is important to ensure that both heat-pump applications and policies are based on a careful assessment of the facts and that the successes and energy-savings information are made available to the public.

So how do these energy-efficient machines work? Most heat pumps work on the principle of a *vapor-compression cycle*. The main components in the heat-pump system are the compressor, the expansion valve, the evaporator, and the condenser. The evaporator section and the condenser section are actually two heat exchangers. The components are connected to form a closed circuit, as shown in Fig. 8.9. A compressible, volatile liquid, known as the *working fluid* or *refrigerant*, circulates through the four heat-pump sections. In the evaporator, the temperature of the working fluid is kept lower than the temperature of the heat source, causing heat to flow from the heat source to the working fluid, and the working fluid evaporates. The compressor is designed to compress, or increase the density of, a compressible fluid or refrigerant.

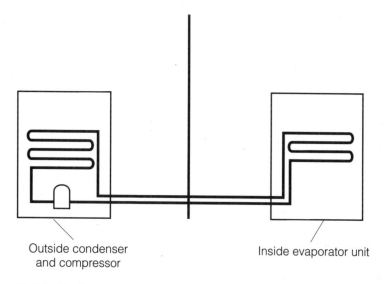

Outside condenser
and compressor

Inside evaporator unit

■ **8.9** *The heat pump's closed circuit.*

Vapor from the evaporator is compressed to a higher pressure and temperature. The hot vapor then enters the condenser, where it condenses and gives off useful heat. The condenser is designed to condense the refrigerant in any air-to-refrigerant or water-to-refrigerant heat-exchange system or as part of a vapor-compression or absorption refrigeration cycle. Finally, the high-pressure working fluid is expanded to the evaporator pressure and temperature in the expansion valve. The working fluid is returned to its original state, and once again enters the evaporator. The compressor is usually driven by a high-efficiency electric motor with very low energy losses. The overall energy efficiency of the heat pump strongly depends on the efficiency by which the electricity is generated.

Another type of heat pump works on the principle of absorption. While this type is rarely found in residential applications, commercial, school, and hospital systems have been employed. Absorption heat pumps are thermally driven, which means that heat rather than mechanical energy is supplied to drive the cycle. Absorption heat pumps for space conditioning are often gas-fired, whereas industrial installations are usually driven by high-pressure steam or waste heat. Absorption systems use the ability of liquids or salts to absorb the vapor of the working fluid. The most common working pairs for absorption systems are water as the working fluid and lithium bromide as the absorbent or ammonia as the working fluid and water as the absorbent. In absorption systems, compression of the working fluid is achieved thermally in a solution circuit that consists of an absorber, a solution pump, a generator, and an expansion valve. Low-pressure vapor from the evaporator is absorbed in the absorbent. This process generates heat. The solution is pumped to high pressure and then enters the generator, where the working fluid is boiled off with an external heat supply at a high temperature. The working fluid (vapor) is condensed in the condenser while the absorbent is returned to the absorber via the expansion valve. Heat is extracted from the heat source in the evaporator. Useful heat is given off at medium temperature in the condenser and in the absorber. In the generator, high-temperature heat is supplied to run the process. A small amount of electricity may be needed to operate the solution pump.

Heat flows naturally from a higher to a lower temperature. Heat pumps, however, are able to force the heat to flow in the other direction using a relatively small amount of high-quality drive energy (electricity, fuel, or high-temperature waste heat). Thus heat pumps can transfer heat from natural heat sources in the surroundings, such as the air, ground, or water, or from human-made heat sources such as industrial or domestic waste to a building or an industrial application. Heat pumps also can be used for cooling. Heat is then transferred in the opposite direction, from the application that is cooled to surroundings at a higher temperature. Sometimes the excess heat from cooling is used to meet a simultaneous heat demand. In order to transport heat from a heat source to a heat sink, external energy is needed to drive the heat pump. Theoretically, the total heat delivered by the heat pump is equal to the heat extracted from the heat source plus the amount of drive energy supplied. Electrically driven heat pumps for heating buildings typically supply 100 kWh of heat with just 20 to 40 kWh of electricity. Many industrial heat pumps can achieve even higher performance and supply the same amount of heat with only 3 to 10 kWh of electricity. The heat delivered by a heat pump is theoretically the sum of the heat extracted from the heat source and the energy needed to drive the cycle. The steady-state performance of an electric compression heat pump at a given set of temperature conditions is referred to as the *coefficient of performance* (COP). The COP is an efficiency measure for cooling source equipment. It is the relationship of the actual cooling effect, or output, to the energy input. The cooling effect and energy input must be in similar units. COP also can be defined as the ratio of heat delivered by the heat pump to the electricity supplied to the compressor. The operating performance of an electric heat pump over the season is called the *seasonal performance factor* (SPF). It is defined as the ratio of the heat delivered to the total energy supplied over the season. It takes into account the variable heating and/or cooling demands and the variable heat source and sink temperatures over the year and includes the energy demand, for example, for defrosting. A heat pump actually defrosts itself. During the defrost function, the heat pump's condenser appears to be stalled. Actually, the condenser fan is not turning, and there is electrical energy being forced to the condenser to "defrost" the components whenever it is very cold outside. Once done, the condenser fan comes on, and the defrosted moisture blasts out of the condenser in a large cloud.

The actual performance of heat pumps is affected by a number of factors, and every installation will be different. For residential heat pumps, the climate (annual heating and cooling demand and maximum peak loads), the temperatures of the heat source and heat distribution system, the auxiliary energy consumption (pumps, fans, electric coils, etc.), the sizing of the heat pump and the operating characteristics of the heat pump, and the heat-pump control system all play major roles in its efficiency and success.

Gas-Pack Systems

A gas pack uses either natural gas or propane gas for heating, and it incorporates an electric air conditioner for cooling. All equipment needed is packaged together with the gas pack; thus it takes on the appearance of a specialized device. Rather, it is a dual system that can supply both warm and cool air to a house through a duct system. While a furnace must have a split air conditioner added to provide cooling, a gas pack includes both in the same unit. The gas pack, like the heat pump, has all the components to provide heating and cooling in a single unit. Package units are located outside your home, are weatherproof, and usually are placed right up against the side of your house. Ductwork is run from the unit (at the side of your house) under the floor in the usual manner.

Following is a set of typical specifications for a single-package gas-pack system:

Up to 14.5 SEER (An air conditioner with a SEER rating of 12 or higher is considered to be highly efficient.)

90.4 percent AFUE (The higher the percentage, the more efficiently the gas- or oil-fired equipment uses fuel. A rating of 90 percent or higher is considered highly efficient.)

High-efficiency gas heating and electric cooling system in one compact, outdoor package

Up to 50 percent energy savings year-round without taking up valuable indoor space

5-year limited warranty on the compressor

10-year limited heat-exchanger warranty

1-year limited parts warranty

Natural gas heating/cooling

Tested in conjunction with the American Gas Research Institute

Built-in humidity control

Driven by a natural gas engine

Fully programmable 7-day thermostat with microprocessor controls

Geothermal Systems

Geothermal systems, sometimes called *water source, ground loop,* or *earth-coupled heat pumps,* use water to transfer heat to and from a house. Some open-loop systems, nicknamed "pump and dump," transfer heat directly to well water and "dump" the water immediately. This water is not used again, and these systems are not as common as the closed-loop systems. The closed-loop systems use a loop of plastic pipe, filled with water, buried in the ground to transfer the heat in and out of your home. These systems are more expensive to install than an air-source system but often provide improved performance during the heating season. The harshness of some winter days will have little effect on the heat-producing capability of a closed-loop geothermal system because the ground temperature below grade is fairly constant.

A geothermal heat pump is powered electrically and pulls heat from the earth, where the temperature remains constant, at 54 to 58°F (Fig. 8.10). The heat is extracted by means of a buried vertical, horizontal, or closed loop of polyethylene pipe that circulates a water-antifreeze solution. This solution collects heat from the earth and carries it through the system and into the house. In the hot summer, the same system pulls heat from the home and transfers it to the earth, providing an efficient way to air-condition. Geothermal heat pumps can cut home heating costs as much as 70 percent in the winter, can reduce home cooling costs up to 25 percent in the summer, and can provide hot water for normal household use.

Because they are self-contained systems, geothermal heat pumps can be installed completely indoors. Since they are inside, they are sheltered from the extreme outside weather conditions that conventional systems must endure. The system is basically quiet because there is no outside con-

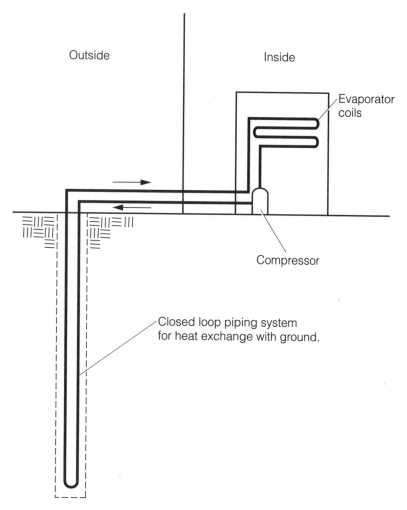

Outside

Inside

Evaporator coils

Compressor

Closed loop piping system for heat exchange with ground.

■ **8.10** *A geothermal heat-pump system.*

denser unit with a fan and rattling parts to disturb the neighbors. Even though a geothermal heat-pump system may cost more to install initially, it normally will carry a longer life expectancy because there are fewer moving parts. With fewer parts subject to breakdown, maintenance is kept to a minimum.

Electrostatic and Electronic Air Filters

Typically, electrostatic and electronic air filters are not designed into a residence upfront. If the ventilation scheme is sound and the indoor air quality is good, then there is no reason to spend the money. However, frequently, dust and particulates that were not predicted become a nuisance and a health problem. Then an add-on filter must be considered. Both electrostatic and electronic filters remove airborne particles when contaminated air is passed over an electrically charged filter. The charge is provided externally from normal household electric current in an electronic filter. In an electrostatic filter, the flow of air causes the filter to charge. It is the electric charge that attracts airborne contaminants to the filter. From a maintenance standpoint, both filters must be cleaned periodically. Also, it should be noted that electrostatic filters have much greater resistance to airflow than electronic filters. Thus, if you are considering an electrostatic filter, ensure that the return duct is airtight, or else there is a risk of increased duct leakage.

Filtration, not air infiltration, is the method by which particulates and sometimes gaseous pollutants may be removed from the air. Pollutants are intercepted by a filter that allows clean air to pass through. This method of air cleaning is necessary when high concentrations of particulates are present or when the polluting source is derived from outside the building. Potential benefits can include improved air quality, reduced dependence on ventilation, and improved energy efficiency. Filtration is not a substitute for the ventilation needed to meet the good-health-sustaining requirements of occupants.

Humidifiers and Dehumidifiers

How often have we heard the expression, "It's not the heat, it's the humidity!" While we hear this often, it is also very true. Temperature, especially the ambient temperature, which does or does not make us comfortable, is directly affected by the amount of water vapor in a given amount of air. *Humidity* is defined simply as the amount of water vapor in the air. For a given volume of air at a certain temperature, there is a maximum amount of water vapor that it can hold until it is completely saturated. This is the *relative humidity* of the air. Saturated air would have a relative humidity of 100 percent. To the home owner, relative humidity is a very important factor in trying to achieve energy savings.

First, we must realize that there are two media to be concerned with when it comes to humidity. The air outside the house has a water vapor content that is beyond our control. It changes with the seasons and can be high or low depending on the region, weather changes, and time of day. What is controllable, when it comes to outside air, is the amount that is let indoors. The air inside the house is that which we are trying to maintain and condition. If we allow great amounts of humid outside air into the house, the air-conditioning system must work harder (thus consuming more energy) to remove the moisture. The indoor air already has moisture in it from the human occupants, showers, cooking, and so on. Extra outside air only aggravates the situation.

As was mentioned earlier, air at 100 percent relative humidity is fully saturated, and condensation can occur quite rapidly. This obviously is not a good condition for a house's interior and the structure itself. The windows, doors, and wooden components can start to show signs of mildew because of the high moisture content. A house typically will have a lower relative humidity in the winter (drier air) than in the summer. This can be measured with a hygrometer. If the house has a relative humidity over 50 percent continuously, then the air-conditioning system is using more energy to remove that moisture, and long-term damage is being done to the house.

One solution is to remove the moisture with a dedicated dehumidifier. The dehumidifier will require less energy to run than the whole-house air-conditioning system. It can be installed as an integral part of the system or can be a remote, self-contained unit. The dehumidifier should operate on automatic, humidistat control, turning on and off whenever the moisture level exceeds the setpoint. Additionally, there should be a nearby drain for the water that is collected. In this manner, the moisture content of the air can be held in check. Of course, if doors and windows can be kept closed during times of high outside humidity, then the introduction of outside air to the house's interior can be reduced. A hygrometer that can be placed so that it displays the outside relative humidity in association with an internal hygrometer can tell the home owner exactly what the conditions are so that appropriate measures can be taken.

Just as air with too much moisture in it is not good for a house, air that is too dry is not ideal either. Very dry interior air can be unhealthy and can dry out wooden doors and windows. In such cases, a humidifier, a device that automatically introduces small amounts of water vapor into the house air, is in order. While dry air is not as common a problem as air that is too humid, there are regions where the ambient outside air is very dry year-round. In these regions, an add-on humidifier is often a solution.

Water Heaters

The most common method of heating water for domestic use is by either gas or electricity. Electric water heaters use resistance elements located within a storage tank to transfer heat directly to the surrounding water. In hot water tanks where two elements are used, the lower element is used to maintain the desired water temperature, while the upper element will energize only when demand for hot water increases. When the water at the top of the tank reaches the desired temperature, the upper element shuts off, and the lower element will turn back on. In this way, the top water in the tank will not get too hot. Each element has its own independent thermostat for setting the desired water temperature. These are typically not adjustable by the home owner and are factory set.

A gas water heater uses a burner located near the bottom of the storage tank. Air combines with the gas that is piped to the water heater, and the mixture is ignited. Heat is transferred to the water through the metal surface of the tank. A vent pipe is used to carry the combustion gases to the outside. The vent pipe is also a heat exchanger that transfers the heat to the water and is usually surrounded by the tank. Baffles are used inside the vent pipe to slow down the combustion gases so that more heat can be transferred to the water around the pipe. Most gas water heaters use a single thermostat for setting the desired water temperature.

Most water heaters, whether electric or gas, have similar water piping and draw hot water from the top of the tank while supplying cold water near the bottom of the tank. While storage tank systems are the most common, "instantaneous" and "tankless" systems are available. If you have special hot water needs, such as large hot tubs, proper sizing of your tank is essential. Heat-pump water heaters, heat-recovery units, and solar water heaters are also available. Consult a licensed plumber before considering these specialized options.

The differences in operating costs of electric and gas water heaters depend on many factors—electric and gas rates, lifestyle and use patterns, specific appliance requirements, etc. A general rule of thumb is that the more hot water you use, the more a gas water heater is the better choice. In most households, once you get to the three-person level or have a large hot tub that is used frequently, gas water heating is probably the best choice. However, for average to small users of hot water, electric water heaters can be the least expensive option. Many electric companies offer incentive programs for electric hot water heater installations. Frequently, the larger the electric hot water tank capacity (60 gallon and up), the better the incentive. Sometimes a monthly rebate is given whenever larger tanks are used. Additionally, residential customers often can finance an electric water heater at a decent interest rate through the electric company. Likewise, gas hot water heaters can come with complete installation, incentives, and financing options for qualified customers in homes where electric water heaters are currently used. There is financing available for gas-to-gas changeouts, but no rebates are awarded. There are also incentives to install natural gas water heaters in new homes.

A hot water tank should last from 15 to 20 years. This is with good maintenance and good water quality. There will be a periodic replacement of elements. Occasionally, you may have to replace a thermostat on the element.

In some cases, water heaters have been known to last more than 30 years. The hours of operation, maintenance, and quality of the supply water will affect the life of any water heater. If you are experiencing early water heater failure, an inspection by a licensed plumber may be a good idea. Many different models of water heaters are available, and some of these may be better suited to your particular needs or application. Most hot water tanks come with insulation wrap to help keep the efficiency and energy ratings high. Another step that can be taken is to insert a piece of polystyrene insulation underneath the tank, as can be seen in Fig. 8.11. This provides a thermal break between the cooler concrete floor and the tank itself. It minimizes any heat losses that could occur through conduction.

■ **8.11** *Electric hot water tank with insulation between tank and concrete floor.* (Courtesy of Wayne J. Henchar Custom Homes.)

Ductwork Systems

After the air has been conditioned, it has to be delivered to the locations within the house as designed. Diffusing units, registers, and grilles have been located to give the proper circulation and dispersion throughout the occupied space. The conveyance system to perform this function is the duct system, or ductwork. This function is the supply air network, and once delivered, this same air must be brought back to the air-conditioning system to repeat the cycle of conditioning. The delivery system is called the *supply system*, whereas the air is brought back to the conditioner by the *return system*. While both are necessary in any ductwork system, the supply ductwork is often more complex and more detailed. Return air is allowed to "float around" for awhile before it is brought back to the air conditioner, whereas the newly conditioned air must be delivered as quickly as possible to maximize the energy consumed while conditioning.

Ducts are usually made of metal, either aluminum or galvanized steel sheets. Residential ductwork typically is of a light gauge, or thickness, and most often the ducts are rectangular. Larger, flat panels of sheet metal used in ductwork usually are crimped diagonally to minimize rattle and for strength. Since the mid-1940s, fiberglass insulation has lined the interior of metal ducts for acoustical and thermal purposes. The registers and grilles are designed for effective air distribution within spaces to be conditioned. Their size, shape, and style are all critical to effective air distribution. The distinction between a grille and a register is in the damper. A damper is part of a register assembly allowing mechanized control of the delivered air at the register. A register is adjustable, whereas a grille is not.

Location of ductwork is determined in the design stages, and once installed, ductwork is not easily relocated. Locating ductwork in the attic will lead to more heat gain in the summer months. Generally, it is better to locate ductwork in the crawl space because this will help with cooling loads. The importance of a well-sealed ductwork system cannot be stressed enough. Completely sealed ductwork is critical to the economical and efficient operation of a heating and cooling system. A leaky return ductwork system will tend to pull outside air into the ductwork, thus pressurizing the house. This will cause excess infiltration, increase inside humidity in the summer, dry the house out in the winter, decrease comfort, and increase operating costs, none of which is attractive to the home owner. The same can be said for leaky supply ducts, except that these leaks will tend to depressurize the house.

Air Changes

Air changes in residential energy systems are the biggest contributor to heat loss. If heated or cooled air finds a quick path out of the house due to infiltration and drafty construction, then energy used to condition that air has been totally wasted. Typically, when calculating the overall heat loss for a residence, the number of air changes will have the most dramatic effect on total loss. Control of air changes, proper ventilation, and control of infiltration play a critical role in energy savings.

An *air change* is defined as the number of times per hour that a volume of air equal to room volume is replaced. One air change per hour, sometimes abbreviated as ACH, represents the supply and return of air equal to the volume of a room or structure once every hour. This is illustrated in Fig. 8.12. The approach of today's home builders is to make the residence extremely airtight. This is correct in theory but not particularly practical. Granted, according to Table 8.2, too many air changes per hour can take its

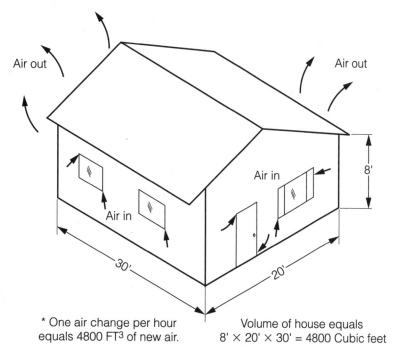

* One air change per hour equals 4800 FT³ of new air.

Volume of house equals 8' × 20' × 30' = 4800 Cubic feet

■ **8.12** *The effect of one complete air change.*

■ **Table 8.2 Air Changes**

Room Type	Number of Air Changes per Hour
Room with no windows or exterior doors	½
Room with windows or exterior doors on one side	1
Room with windows or exterior doors on two sides	1½
Room with windows or exterior doors on three sides	2
Entrance hall	2

toll on a heating or cooling system. However, some fresh air has got to be allowed into the house for health reasons, moisture control, and to maintain good indoor air quality. The design scenario should be to select a good balance between losses and necessary fresh air. Proper ventilation also can provide an adequate, energy-efficient solution.

Ventilation

Ventilation is the process by which outdoor or cleaned air is brought into a building or room as a means of maintaining an acceptable indoor air quality. It is also used to control building pressurization and to provide thermal comfort. Often, the ventilation system is required to accomplish tasks for which it is not intended. Typical problems such as humidity control, heating, and cooling are not really the job of the ventilation system, although many systems combine the processes. Outside air, or *makeup air*, is brought into a building to compensate for indoor air that must be exhausted. If too much indoor air is exhausted, then the building can begin to exhibit a negative pressure, seen in Fig. 8.13. Building ventilation schemes and sizing are based on a balanced supply and exhaust design. Obviously, as the rate of ventilation increases, the energy demands rise accordingly. It is not very energy efficient to exhaust great amounts of heated or cooled air.

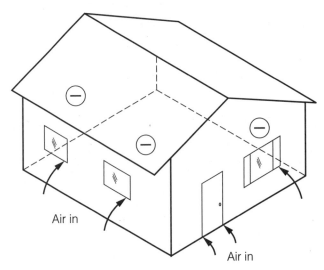

Air in

Air in

Negative pressure inside the
house pulls outside air in through
all cracks and openings.

■ **8.13** *Negative pressure within a house.*

Air is a mixture of gases: oxygen, nitrogen, argon, carbon dioxide, and water vapor, along with any suspended solid and liquid materials (vapors, dust, particulates, etc.). Air, for a given volume, also will contain a certain amount of heat depending on what energy system has acted on it. Thus cleaning and maintaining the air and the heating and cooling of it are not the simplest of tasks. The importance of a well-sealed duct system cannot be overemphasized. This is critical to the economical operation of your heating and cooling system. A leaky return duct will pull outside air into the ductwork, pressurizing the house. This will cause excess infiltration, increase inside humidity in the summer, and dry the house out in the winter, decreasing comfort and increasing energy and operating costs. The same can be said for leaky supply ducts, except that these leaks will depressurize the house.

The buildup of pollutants is reduced by ventilation. *Indoor air quality* (IAQ) is the state or condition of the air in a given space. IAQ is a measure of detectable odors and nondetectable substances that may affect the health of occupants. Ventilation is needed to supply oxygen and control odors within the home. Good indoor air quality may be defined as air that is free of pollutants that cause irritation, discomfort, or ill health to occupants. Thermal conditions and relative humidity also influence comfort and health. A poor indoor environment can manifest itself as a "sick" building in which some occupants experience mild illness symptoms during periods of occupancy. More serious pollutant problems may result in long-term and permanent ill-health effects. Since much time is spent inside buildings, considerable effort has been focused on methods to achieve an optimal indoor environment, with particular emphasis on health, odor control, thermal comfort, and energy efficiency. Indoor air-quality measures place emphasis on the following:

1. *Health of the occupants.* Concentration of contaminants can lead to sickness of the inhabitants. Proper ventilation reduces this risk.

2. *Irritable odor.* Poorly ventilated homes can retain odors and smells that may not be harmful to one's health but are embarrassing and irritating.

3. *Building comfort.* A drafty residence is typically not comfortable, and energy is wasted when ventilation is too great.

4. *Oxygen supply and metabolism.* All humans need to breathe air that contains a certain level of oxygen. When this level drops, the occupants can get sick.

A tremendous amount of the energy consumed in residences, either by heating or by cooling, is lost by ventilation and air infiltration. This is important, obviously, to the consumer, who has to pay the monthly costs, but the world's energy needs and environmental concerns also continue to grow. Moreover, ventilation is closely linked to indoor air quality. Thus, how much ventilation is needed to provide for a healthy indoor environment and still use energy wisely? It is difficult to assess the energy impact of ventilation, and the true number of air changes in relation to energy use is often undefined. As a result, it is rather difficult to know if too much or too little ventilation is adequate whenever designing a system. This difficulty stems from the enormous complexity of the task, which needs to accommodate wide variations in factors such as human comfort and health-need differences, climate, the airtightness of the residence, occupancy habits, and varying approaches to ventilation. A ventilation system must be designed to satisfy the required demand. In meeting this demand, it is often necessary to consider the heating and cooling system approach for the residence, local and state building regulations, and maintenance and replacement issues. It is also necessary to integrate the ventilation system into the overall design of the residence, especially in relation to airtightness, room partitioning, ductwork systems/dampers, the introduction of ac variable-frequency drives, and accessibility. Therefore, ventilation needs must be based on criteria that are established at the design stage of the home. This is where the home owner, home builder, and mechanical contractor must agree. To return after installation in an attempt to correct problems as they arise will lead to considerable expense, construction delays, and long-lasting unhappiness. If the home owner, as the occupant, lives and breathes in the new home and feels that he or she is getting sick from the inadequate ventilation, then the unhappiness continues to compound.

Ventilation should not be taken lightly, especially when the residence will be occupied by older people or anyone with respiratory difficulties. There are many systems and techniques available to meet the needs of ventilation, and each has its own set of advantages and disadvantages. Sometimes the local climate will dictate the ventilation scheme and overall building type. All too often price becomes an issue, and to justify a complex strategy, the home builder must demonstrate advantages in terms of improved indoor climate, reduced energy demand, and an acceptable payback period. A great deal of energy is lost from a residence through the departing air stream. Whenever infiltration is the biggest factor affecting air change, not much can be done to recapture this lost energy. This is why more attention during the design stage must be paid to air infiltration and the actual ventilation scheme.

Cooling is needed when the indoor environment becomes excessively hot or humid. This may occur as a result of high outdoor temperatures or as a consequence of excessive solar or internal heat gains. High internal gain is particularly a problem in large, nondomestic buildings. When the need for cooling is dictated by internal heat gains rather than outside temperature and humidity, much can be accomplished to reduce the need for or eliminate altogether active cooling systems. Solutions depend on climate but include cooling by ventilation (passive cooling), designing for reduced solar gains, the use of thermal mass, and restricting internal heat loads.

Ventilation even gets its own efficiency ratings, that of the distribution of pollutants within the residence and the mixing elements of the air itself. These two aspects may be subdivided into air-change efficiency and the effectiveness of pollutant removal. Ventilation efficiency is based on an evaluation of the age of the air and on the concentration of pollutants within that air. Some ratings are based on room-averaged values, whereas others refer to specific points or locations. This has important consequences because while room values provide some guidance to the overall performance of a ventilation system, point values indicate areas where poor ventilation might occur. This will then allow for corrective measures to be taken.

Maintenance is an issue to cover regarding the reliability of the ventilation system. The long-term success of any ventilation system will be in its ease of maintenance and its proper installation. Maintenance is often overlooked, and the need for maintenance may even be ignored in the course of the residence's design. Typical problems include worn gaskets, dirty fans and grilles, and ill-fitting and clogged filters. Not many home owners routinely change their air filters, but they should. The system will run more efficiently if they do. This issue has resulted in an attempt to provide more "maintenance-free" system whenever possible. Only by correct functioning can a ventilation system be relied on to meet the indoor air quality needs of the residence. Measurements are needed to verify the performance of ventilation systems and to test the integrity of the building shell. They are essential for commissioning, diagnostic analyses, design evaluation, and research. In addition, measurement results provide the fundamental means for understanding the mechanics of ventilation and airflow in buildings. Measurement data are also needed to provide background information for parametric studies on building air leakage characteristics, indoor air quality, and ventilation system performance. Many measurement techniques have been developed, and each has a specific purpose.

To some, ventilation may be a very minute issue when it comes to house design, whereas to others, it is very important. In either case, typical ventilation designs must cover system sizing, performance evaluation, indoor air quality prediction, energy impact assessment, and cost-benefit analysis. Frequently, a model is used to verify the calculations. Such a process is helpful, with adjustments made to parameters over which control is possible, until an optimal design solution is achieved. A large number of methods of varying complexity have been developed, with no single method being universally appropriate. Selection of proper equipment varies according to the required level of accuracy, the availability of data, the needs of the home owner, and the type of structure under consideration. As designs have become more complex and performance tolerances more demanding, it is increasingly important for the designer to be able to understand and use calculation techniques. This need has resulted in the development of improved algorithms and wider availability of design data.

Whole-House Fans

We have seen in many instances where it is important to keep air moving within a residence. Air movement lessens the demand for cooling and, ultimately, the need to consume more energy to run the cooling system through a cycle. Additionally, the movement of air can introduce a small amount of fresh air into the residence, thus not allowing stagnant air to accumulate. Stagnant air can be a medium for unhealthy bacteria to grow. The energy that a whole-house fan uses is miniscule when compared with the amount consumed by a heat pump through several cycles over the same period.

The concept of a whole-house fan is not new and has a few variations. Evening air is usually cooler than daytime air during the hot summer months. Blasting the warm out and pulling in fresh, new, cooler air is the whole-house fan's primary function. If the outside temperature is at or higher than the inside temperature, then the whole-house fan will only reposition warm air, thus not providing any real cooling. Additionally, the humidity of the outside air should be considered, because hot and humid air being brought into a house will be both uncomfortable and make the cooling system work even harder. Many homes are constructed with attic fans that are thermostatically controlled to move hot air out of the attic during the summer months (Fig. 8.14). These fans do not introduce any new air into the general living quarters of the house but rather are smaller fans that keep the plenums from building up heat.

■ **8.14** *Thermostatically controlled attic fan.* (Courtesy of Wayne J. Henchar Custom Homes.)

There are some disadvantages to whole-house fan systems. One drawback is that they are typically noisy, and the bigger they are, the noisier they are. Thus, if there is a need to evacuate a lot of air from the house quickly and the house is somewhat large, the fan could be large. Some fan motors could approach 5 hp or even 10 hp in size, and therefore, more wattage and current will be used. The motor size is directly related to the cubic feet per minute (ft^3/min, or cfm) of air is required. This is a common unit of air volume used in the heating, ventilating, and air-conditioning industry. The higher the rating in cubic feet per minute, the more energy will be necessary to operate the fan. This must be weighed against the power needed to do the cooling. Whole-house fans also need shutters and dampers so as to keep cold air outside during the winter and hot air outside during the summer. These devices typically operate (open) when the fan is energized, often by the pressure that the fan exerts. Sometimes these shutters and dampers get stuck in the open position and have to be forced shut to keep from losing inside air and allowing outside air in whenever such conditions are not desired. In addition, shutters and dampers leak small amounts of air when not in use. Even the best of seals around these devices still permit some air movement.

Centrifugal Pumps

A centrifugal pump is a system unto itself. A *pump* is defined as a device that imparts energy (velocity and/or pressure) to water in an HVAC system. Pumps are commonly found in chilled water, hot water, and condenser water circuits. They are centrifugal in operation because they are a particular type of fluid-moving device that imparts energy to the fluid by high-velocity rotary motion through a channel. The fluids enter the device along one axis and exit along another axis. Pumps are made up of an ac electric motor, an impeller with housing, various seals, intake and discharge sections, and if we include the piping to and from the pump, much more. In the home, there are several pumps at work. They fill and empty the washing machine and dishwasher and for many provide the potable water we use everyday. There are also pumps at work on our swimming pool filters, health spa or whirlpool bathtubs, oil furnaces, and so on. Thus such pump systems really are individual energy systems, each consuming energy while performing work.

Most radial-flow, or centrifugal, pumps have similar components and operate in much the same manner. The centrifugal pump generally moves high volumes of fluid at low pressures. If we look at a typical speed-versus-torque curve for a centrifugal pump, we find that at low speeds, not much energy is required because the loading is light, but as we continue to increase in speed, there is more flow of the liquid, meaning we need more motor torque to continue moving the load. This is typical of virtually all the small pumps within a house. However, the ac motor is starting with high inrush current, and the current draw is high throughout the time it takes to get to full speed (and at this point we still need full load current because the flow demands it). By placing an ac variable-frequency drive on this motor, the current supplied to the motor is only the amount needed to perform the work, or flow, at that part of the cycle. Thus full current is not delivered until the motor is close to full speed. Additionally, an ac variable-frequency drive placed on a pump motor to reduce the motor speed will reduce the flow when lesser demand is all that is needed, and energy can be saved. Chapter 9 looks at variable-frequency drives in greater detail.

Because the pump system must have the capability to always run (even if the drive is faulted or fails), a line bypass method is usually installed with the drive (on ac systems only). A partial vacuum is formed when the centrifugal pump impeller rotates, thus drawing more fluid into the pump. However, when the suction is broken, cavitation takes place. These load interruptions could trip the drive off-line. Likewise, "hammer" in the piping system can exist, sometimes when a centrifugal pump operates under a lot of head pressure. This is not so much a problem for the drive as it is for the piping system itself. The obvious benefit is the energy savings. Instead of using valves and vanes in the system, the ac drive controls the flow. Another benefit is the ability to run full speed via the bypass system. Yet another benefit is soft acceleration and soft deceleration; this can help reduce hammer in a pump's piping system. The ac drive also can "ride through" many line fluctuations and through some instances of pump cavitation. The electronic ac drives provide energy savings and the ability to bring a pump motor up from a soft start to full speed and constantly protects the motor from overload conditions.

Piping and Plumbing Systems

Fortunately, the piping and plumbing systems are not usually a major factor in heat loss in a residence. While attention should be paid to hot and chilled water pipes (wrapping them with suitable insulation) so that energy losses are kept to a minimum, the piping system is the transport vehicle for the overall heating or cooling system in the house. While a piping system's energy-savings contribution will remain low, it is still very important to select piping materials carefully. A pipe, the container for distribution of water, gas, refrigerant, or steam, can come in many material types and many different sizes. Some materials are better than others for hot water, wherease others are easier to work with on installation. Table 8.3 shows various piping materials, connection means, and notes on each.

Some piping will be better suited for certain applications based on pressure and temperature. *Pressure*, a measure of the force exerted by a fluid, is not usually an issue with water supply and waste water systems but is a consideration with steam and refrigerant piping systems. *Static pressure* is a measure of the force exerted by a fluid at rest, whereas *velocity pressure* is a measure of the force exerted by the momentum of a fluid in motion. It should be noted that total pressure, relative to piping, is the sum of static and velocity pressures.

■ **Table 8.3 Various Piping Materials**

Kind of Pipe	Material or Manufacture	Connections	Qualities	Notes
ABS	Acrilylonitrile-butadiene styrene	Epoxy cement	Corrosion-resistant	
Brass, red	85% copper, 15% zinc	Threaded, IPS (iron pipe size)	Corrosion-resistant	Bulky because of the need for threading
Copper tube, type K	Seamless, hard or soft temper	Soldered fittings	Corrosion-resistant and easy to fabricate	Thinner-walled than brass; easy to put together and dismantle
Copper tube, type L	Seamless, thinner walls than type K, hard or soft temper	Soldered fittings	Corrosion-resistant and easy to fabricate	Thinner-walled than brass; easy to put together and dismantle
Galvanized steel	Zinc-coated steel	Threaded	Moderately corrosion-resistant	Suitable for mildly acid waters
Nickel, silver, and chrome	Copper, nickel, and zinc, steel, and chromium	Threaded	Corrosion-resistant	Special applications
PE	Polyethylene	Epoxy cement	Corrosion-resistant	
PVC	Polyvinyl chloride	Epoxy cement	Corrosion-resistant	
PVDC	Polyvinyl dichloride	Epoxy cement	Corrosion-resistant	
Steel	Butt welded to 2-in-diameter seamless, large sizes	Threaded	Basic	Should be used only when water is not corrosive
Wrought iron	Butt welded to 2-in-diameter seamless, large sizes	Threaded	More corrosion-resistant than steel	Identified by a red spiral stripe

Ease of installation and initial costs are the other factors in pipe selection. The home builder and the mechanical contractor should meet with the home owner to discuss pipe routing, valving, and future access. Valves do not last forever and eventually must be replaced. Having an access door or panel when the time comes to replace a valve is far better than having to tear apart a wall and then rebuild and refinish it. *Valves*, devices designed to control water flow in a distribution system, are the wear items in the piping circuit and can contribute to energy losses whenever they corrode and do not function properly. Common valve types include check valves, globe valves, gate valves, and butterfly valves.

Motorized Shutters and Awnings

As we continue to strive to collect as much solar energy in a day as possible, there are times during the summer months that we do not necessarily want that extra heat. As a matter of fact, the extra heat makes the cooling system work harder, thus consuming more energy. Therefore, a few built-in mechanisms are necessary by which we can account for seasonal changes and *not* cause our efficient cooling system to work harder. One method used to keep the sun's rays out during the summer months is to cover the windows, pull the shades, or use shutters. These are basically manual methods, whereas motorized shutters can be used in their place. Motorized shutters can even be linked conveniently to a home automation system to open and close based on room temperature. These are not necessarily the small shutters used in conjunction with whole-house fans, but rather motor-actuated shutters that move over windows to allow sunlight in and keep sunlight out depending on the time of year and climate.

Closing Remarks on Mechanical Systems

As can be seen, the mechanical content of the average home, new or old, is fairly large. Sometimes up to 25 percent of the cost of a home can be apportioned to mechanical systems such as the heating and cooling equipment, the ventilators, hot water heating, refrigeration equipment, plumbing, and so on. It adds up quickly to be a major factor in not only the final price for the residence but also the final design. For these reasons, it can be justified that the mechanical system market is a multibillion dollar market. This is also why close communication and coordination between the home owner, home builder, and even the mechanical subcontractors during the design stage is advised.

Once a mechanical system is incorporated and installed within a residence, it is very difficult to change, remove, or replace. It is extremely critical to discuss the issues concerned with the selection and design of a particular system. The home owners have to make their health, future, and living needs apparent to the home builder during the development of the home in the initial design stages. The home builder can guide and suggest, but the eventual occupants must communicate their needs. Unfortunately, many mechanical systems are misunderstood by many home owners as to function, operation, and apparent need within the home. This is where the home builder must help the home owner.

Residential Electrical Systems

9

Isn't it amazing that so much of our everyday life is affected by electricity in one way or another. Maybe we should pay more attention to it—because so much about it is taken for granted. How does electricity relate to amperage, voltage, and power? What appliances within our homes are "electricity guzzlers," and which are actually energy efficient? What are we actually paying for anyway? In order to attempt to save and conserve expensive electricity, we should have an understanding of what electricity is and how it is created, know the relationships to actual power, and even understand how electricity is delivered to every residence in our area. Beyond this, how is that same electricity delivered to each individual appliance in the home? How many circuits does my house have and what type of service does it have? We need a starting point in order to use our electricity more efficiently. The electricity that enters our homes is a part of the home's total energy system, yet that electricity itself is a system. There are actually many components necessary to deliver and use electricity within a home. This is why we must analyze the residential electrical system. Figure 9.1 shows a typical residential electrical system. This system is a network of electrically driven appliances, lights, and mechanical devices. Every year more electrical products are introduced and added to our home's existing electrical system network. Over time, the home's electric circuits and the actual electrical service itself reach maximum capacity. This trend demands that we have a full and complete understanding of all the electrical products in our homes and the electrical systems themselves.

The Monthly Electric Bill

Every month we receive a bill from the electric company, and we pay it. We may review the telephone bill or charge card bills, and if a call or charge is in error, we usually get it corrected—taken off the bill. How many people really know what's on their monthly electrical bill. Have you ever had any charges removed from your electric bill or even called the electric company

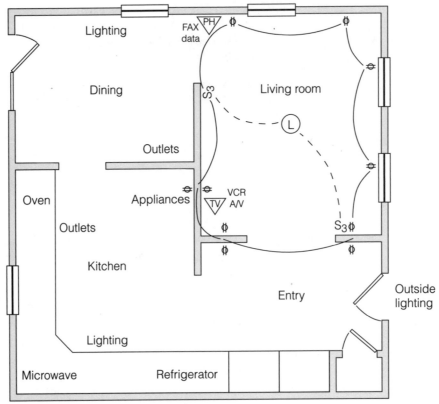

■ **9.1** *Typical residential electrical system.*

at all? How do you know it is accurate? Granted, many home owners read their own meter each month to avoid peaks and valleys in their bill amounts due to the dreaded "estimated" meter reading, but most just receive the bill, look at it quickly, mentally compare the amount to what they perceive was last month's amount, and pay it. No more questions, yet there should be.

How much are we being charged per kilowatthour? What is a kilowatthour anyway? What are all those "other" charges: monthly demand, basic service charge, billing load in kilowatts? Can you buy your electrical power from another source (in the future, the brokering of electricity should become a reality). Additionally, various electric company rates and billings are typically different in some way. In your region, make some inquiries about rates and talk to others who have a different electrical provider. Gain as much insight as you can. Call your electric company and ask specifically how your monthly costs can be reduced. Frequently, the electric company offers special discounts for installing energy-saving equipment. The utilities also offer from time to time rebates for installing energy-saving equipment. They have been known to do this with industrial and commercial users as well as with residential users. Ask about any such discounts being offered in your area.

A sample monthly electric bill is shown in Fig. 9.2 and a sample electric meter display is shown in Fig. 9.3. For this example, they will interrelate. The electric meter is installed at the service entrance to the residence. The meter is the property of the electric company and cannot be tampered with. It operates simply by the electric current running through it; that current which you are eventually going to be billed for actuates tiny gears that turn indicators. This is the reading and is what you pay for as kilowatthours (kWh) used. It is the difference between last month's reading and this month's reading. Also, it should be noticed that there is one large rotating disk that initiates the movement of the geared indicators. If this disk is spinning around at a high rate of speed, then there is probably an electric appliance on such as a heat pump, electric dryer, or some other electricity "hog." The faster that disk spins, the faster your kilowatthours accumulate!

Sample Monthly Electric bill.

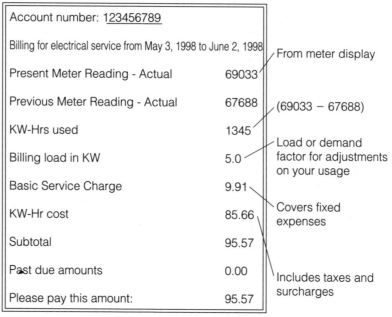

Account number: 123456789	
Billing for electrical service from May 3, 1998 to June 2, 1998	From meter display
Present Meter Reading - Actual	69033
Previous Meter Reading - Actual	67688	(69033 − 67688)
KW-Hrs used	1345
Billing load in KW	5.0	Load or demand factor for adjustments on your usage
Basic Service Charge	9.91
KW-Hr cost	85.66	Covers fixed expenses
Subtotal	95.57
Past due amounts	0.00
Please pay this amount:	95.57	Includes taxes and surcharges

■ **9.2** *Sample monthly electric bill.*

Consider this: Are all electric meters built exactly alike? Could one have more excitation than another? They function based on a mechanical sequence initiated by an electric current. Over time, could you be paying for more accumulated hours? Where and when are electric meters calibrated

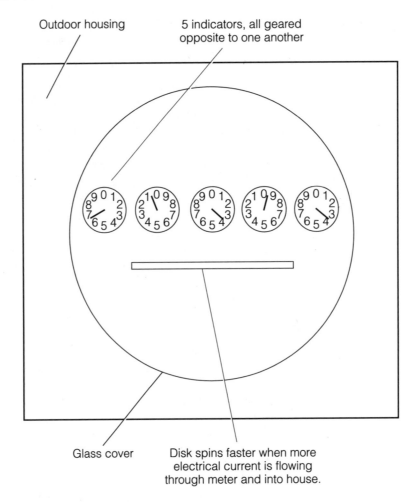

Outdoor housing

5 indicators, all geared
opposite to one another

Glass cover

Disk spins faster when more
electrical current is flowing
through meter and into house.

■ **9.3** *Common outdoor electric, or single-stator, watthour meter.*

(a process of ensuring that the meter works within a specified list of parameters, usually the bureau of measurements and standards governs commercial equipment such as this). Should they be calibrated frequently, as are gasoline pumps or common volt-ohm meters used in industry? How often has the electric company been out to calibrate or certify that the meter is functioning properly over the past 25 years or life of your house?

A *kilowatthour*, sometimes abbreviated kWh, is basically defined as 1000 W of electricity used in 1 hour. It can be derived many ways: one incandescent light, a 100-W bulb, being powered for 10 hours or ten 100-W lights being powered for 1 hour. It all gets accumulated one way or another, and it accumulates quickly! These devices all mainly are powered by 115-V ac, 60-Hz, single-phase power. Many of the 220-V ac appliances (electric range, hot water tank, furnaces, etc.) cause the kilowatthours to increase more rapidly.

The cost per kilowatthour is often not shown on the electric bill but can be calculated from the kilowatthours used and the total cost for these hours. Perhaps the actual cost per kilowatthour is not shown so as to keep the average home owner "in the dark." Rates vary all over the country from lows of 2 cents per kilowatthour to highs of 14 cents per kilowatthour. Obviously, the higher the cost per kilowatthour, the more dramatic is the impact on the monthly electric bill. Often, electric companies have varying rates and load or demand adjustments to give the appearance that the consumer is being given a break, but it would be very difficult to find a home owner or home builder who would agree that electricity is inexpensive! For example, on our sample bill, the preceding meter reading subtracted from the current meter reading gives us a number equal to the kilowatthours used for the given period. Dividing this by the total kilowatthour cost yields a number that gives us a starting point for the actual cost per kilowatthour. Thus, if a 100-W light were on for 10 hours, then the cost for that would be about 8 cents, right? Not necessarily so. Additional demand and load charges, coupled with basic service fees and any other surcharges (tax or otherwise), usually will make that light's use and cost for that amount of time actually higher.

The electric company normally will assess a monthly service charge to the bill to cover fixed expenses. These expenses include billing and invoicing paperwork, having the actual meters read monthly, and the electrical equipment necessary to bring the power to each residence in an area. This amount typically does not change from month to month, but if it does, there should be a disclaimer with the bill advising consumers as to why the amount has changed. Understanding all elements of the electric bill will allow you to notice changes more readily. Whenever a surcharge or premium is assessed to any bill, it is wise to know exactly what the amounts are, what they are for, and why they are in place. If you do not ask and do not know, then your money is being tendered fruitlessly.

Another factor that affects the monthly electricity bill is the maximum load or demand. Certain times during the month there is a maximum, or peak, amount of load required by the residence. This could be whenever the air conditioner, washing machine, dishwasher, lights, hot water tank, and any other appliances are all on at the same time—all demanding electricity, thus dramatically increasing the load. The electric company typically has to be ready for multiple residences requesting this high demand. Thus the monthly bill and the actual kilowatthours used per month are additionally affixed a higher cost due to this "load factor." Electric companies will bill many different ways for this and can even authorize reductions, or discounts, in load factor if energy-efficient hot water tanks and/or heat pumps are used. Check with the electric company in your region to see what load-factor-reduction techniques may be employed to reduce your overall monthly costs. Interesting to note, electric companies all tend to have a portion of the monthly bill that is hard to define. This allows the electric company to charge or surcharge, and thus the consumer may not know what he or she is actually being billed for. Always press the electric company for a full, complete definition of billing elements, and *always* ask what can be done to reduce the overall amount of the bill.

Another area in which the electric company exercises penalties and premiums is the *power factor*. While this phenomenon is more prevalent with industrial and commercial users, the residential customer is going to hear more about it in the future. In any alternating-current (ac) power transmission or distribution system there exists a relationship between the electrical voltage and the current being used. It is calculated as a trigonometric function and is the cosine of the phase angle between the voltage and the current. When the load is inductive, e.g., an induction motor, the current lags the applied voltage, and the power factor is said to be a *lagging power factor*. When the load is capacitive, e.g., a synchronous motor or a capacitive network, the current leads the applied voltage, and the power factor is said to be a *leading power factor*. Note that power factors other than unity have deleterious effects on power transmission systems, including excessive transmission losses and reduced system capacity. Power companies therefore require customers, especially those with large loads, to maintain, within specified limits, the power factors of their respective loads or be subject to additional charges.

Monthly electric bills have changed over the years and vary from company to company. State and sales taxes are required to be collected by the electric company, and these usually show up as a surcharge on the bill. However, if these amounts are not displayed as actual amounts, you may inquire as to how much these taxes are. In this manner you will be able to ascertain exactly how many kilowatthours you're using and the amount going to surcharges and taxes and better know which months are more severe than others. Then you can begin to trim some energy costs and get more efficiency out of your home's energy consumption.

Deregulation of Electric Power and Cooperatives

The telecommunications industry was deregulated a few years ago, and so too will be the electricity industry. Government regulation is always a controversial topic of discussion with any industry, but the fact remains today that deregulation of the electric utilities is in motion. The major question is, as always, how will it directly affect me? Your electricity will continue to be delivered from your local utility's poles and wires. However where that power comes from will be the interesting question. Your monthly electric bill will continue to come from your current utility and probably will include new and different charges. However, the whole premise around the brokering of electricity is that the home owner will have the choice to select what power and from what source that power is delivered into the residence.

While not a new concept, government deregulation still takes time—many years and even decades to really take hold. The deregulation of the electric industry has actually been going on for many years. In 1978, the Public Utilities Regulatory Policies Act required utilities to purchase power from unregulated generators. This was to encourage the development of multiple, smaller generating facilities. This act also was designed to force the development of new technologies and alternative fuel sources such as hydroelectric, solar, wind, and waste to produce electricity. Then, in 1992, the federal Energy Policy Act allowed more types of unregulated companies to generate and sell electricity. EPACT, as it has been called, is much more than a deregulating entity. Several standards and guidelines for all energy use are part of the Energy Policy Act.

Four years later, in April of 1996, the Federal Energy Regulatory Commission mandated that electric utilities nationwide allow other electricity providers to transmit electricity through their utility transmission systems. Utilities and other companies in areas where electricity is less costly to produce would be able to sell cheaper electricity to areas where it is more expensive to produce electricity. The northeastern United States and California have been areas of the country where electricity is very expensive not only for the home owner but also for industrial users. Today, Congress is involved with legislation that would make the nation's electricity market more competitive. This will be an ongoing process for years to come. What will the deregulation of the electricity industry and its legislation bring to the average home owner? Only time will tell.

Deregulation is being pushed by large industrial users across the United States. Those areas of the country which have much higher electric rates than are found elsewhere in the United States need some relief. Building new power plants is both too costly and too lengthy a process and thus not a practical option. Rebate programs have been in existence in these areas of the country to promote energy savings and fall in line with the standards set forth by EPACT. However, large industrial users want a radical change in the way electricity is bought and sold so that they can shop for electricity in what may be an open market. This should filter through to the commercial and residential markets as well. However, there are still many questions with varying answers.

What happens in an area of the country where a large industrial user leaves the local electrical system? The costs of providing that particular power are still there, and those costs could be passed on to the home owners. This cost shifting from larger customers to small commercial and residential customers means rates could go up, not down, as predicted. And what about service reliability and support? The electric system in the United States is without a doubt the best in the world in terms of safe, reliable electricity distribution. Deregulation could cause problems by changing this delivery system to operate in ways it was not designed to operate, resulting in power failures and other phenomena. Retail competition in providing electricity and the now deregulated gas market allow for selecting not only lowest cost but also high-quality power, a stable supply, and accurate billing and metering services from the energy provider.

The keys to the success of this deregulation ultimately will lie with the public's acceptance of the program. The typical home owner wants reliable, inexpensive electricity, few or no interruptions to service for any reason, and continuity in the monthly bill (except for price reductions, which they will always accept). The definitions of customers will take on new meaning. There will be *captive* customers, who do not have economically realistic alternatives to buying power from the existing local utility, even if that customer had the legal right to buy from competitors. And there will be *contestable* customers, who will have the opportunity to choose how their electricity needs will be met. They may take their supply from the retail supplier operating in their area or buy from other retail suppliers or from a generator or wholesale pool. Transmission of electricity and distribution to the home will remain a service through the local utility. That local utility will still be responsible for service questions and emergency repairs.

While one should expect some minor changes in the monthly electric bill, there may be instances where other cost elements surface. The price for each kilowatthour will be affected by a source that will display the "spot" price of electricity. Thus the stock market will contain a new element—the power exchange. Monthly electric bills will be itemized so that the home owner can easily see how much he or she is paying for electricity, utility charges for transmission of electricity, any taxation and surcharges, and any competitive transition charges applicable. Current electric rates reimburse utilities for their costs of building power plants and buying electricity from independent power producers. In today's market, this not practical. Additionally, as pools of home owners and groups emerge, their buying power increases. Cooperatives, or coops, will form to provide perhaps the best system to broker electricity and yield the lowest costs to the end user.

Coop is the commonly used term for a rural electric cooperative. Rural electric cooperatives generate and purchase wholesale power, arrange for the transmission of that power, and then distribute the power to serve the demand of rural customers. Coops typically become involved in ancillary services such as energy conservation, load management, and other demand-side management programs to serve their customers at least cost.

On paper, the brokering and deregulation of the electricity industry appear to eventually lead to lower rates to home owners. By definition, an *electricity broker* is a retail agent who buys and sells power for customers or generators without taking any risk or financial position. The agent also may aggregate customers and arrange for transmission, firming, and other ancillary services as needed. It may take some time, but having a choice or two makes for good, solid competitive pricing. Electricity producers and the local delivery companies will have numerous incentives to operate as efficiently as possible so that they can be competitive, remain financially stable, and be forced to show a profit to their shareholders. This all means that the effort will be there to make this electricity-delivery system work. A proficient system should produce lower electric rates over the long haul. Electrical appliance manufacturers will continue to produce products that satisfy consumer needs but use as little electricity as possible.

The larger electric companies will still own their transmission facilities but typically will turn the day-to-day operation of these facilities over to an independent system operator (ISO). This is done often within this industry and the nuclear power industry now. The Federal Energy Regulatory Commission will regulate the operator, so there still will be some government regulation. The operator will operate a state's transmission system to ensure that electricity flowing into it reaches all customers when they need it so that they continue to have reliable service. The operator also will act as a traffic cop, making sure that all generators have equal opportunity to send their electricity through the transmission system to their customers. Generators who ship electricity through the system will pay a fee to cover system costs and to ensure system reliability.

Electricity is a vital energy resource for all. We are more dependent on it today than ever. The cost of electricity becomes a substantial portion of budgets for both industries and businesses (large and small), as well as schools, hospitals, and especially residential consumers. However, with the current delivery system, home owners and other consumers have no choice about where they purchase their electricity. For practically every other product, whether it is a commodity or a service, consumers can make a choice. Home owners can choose what products they want and from whom they'll buy them. However, with today's electricity, there is no choice—we pay a price, established by our government, that is in turn paid to a provider with no competition. Is this a monopoly or what?

Under this traditional regulation, prices for electricity are unreasonably high and vary widely depending on where a customer is located and which utility has the exclusive right to serve that customer. A competitive market for electricity generation at the wholesale level is quickly developing, due to changes in federal and state legislation that have encouraged and fostered this market. New suppliers, such as qualifying facilities, exempt wholesale generators, and power marketers, are entering the market and offering low-cost power and innovative services. However, the only beneficiaries of the new wholesale competition are electric utilities that buy the low-cost power. Retail electricity consumers are still paying rates to monopoly utilities that are significantly higher than the competitive price for power. The most effective way to bring lower prices for power to consumers and to eliminate disparities is to force the utility monopolies to compete for all retail services. Only when all consumers have the opportunity to choose their electricity supplier will utilities begin to lower their prices and provide the services that consumers want.

Once deregulation is in full swing and a new competitive market is in place, home owners will be able to choose how and where they buy their electricity. They will be able to purchase electricity directly from a supplier, or they can use a electrical power broker who can shop for the best price and service for that customer or group of customers. Consumers also can choose to continue receiving service from their local distribution utility. This consumer choice will force all suppliers, including the local distribution utility, to compete to offer low prices and innovative services to satisfy consumer demands.

Competition in the natural gas, airline, and telecommunications industries has led to nonmonopolized and fair-priced markets. Service and support have remained in tact. The same result should be expected within the electrical power industry; electricity prices should drop due to the competition for the power business. Some consumers will be able to stay with their local utility and expect lower prices. If these consumers do not get the lower prices, then they will have the ability to switch, since the utility will understand that the consumer can choose to buy power from another supplier. Not only should prices drop, but the types and quality of service also should improve. Innovative service options have emerged out of the telecommunications and airlines industries due to competition, and the same should be expected from deregulation of the electrical industry. Special electricity service, energy-efficiency programs, and even appliance maintenance (refrigerators, heat pumps, hot water tanks, etc.) will be areas of improved service. New energy technologies could even get more attention, since all providers will be looking for the next advantage. Renewable and "green" power also should rise as a viable alternative.

"Green" Power

What do we mean by the term *"green" power*? "Green" power is an energy source that is renewable. Renewable power is generated from resources that should not run out or are quickly renewed through natural processes. The process of plant photosynthesis allows *green* plants to grow and flourish. Renewable power includes power generated from geothermal, wind, biomass, solar, and even small hydroelectric sources. Renewable electricity technologies are the cleanest, work with the environment, and are being offered in certain locations across the United States. Deregulation of the electrical industry will open up increased competition to provide "green" power. Home owners will have a choice. More and more megawatts of electricity will come from "green" sources. This power will be supplied on the power company's grid, and consumers can elect to use "green" power as their primary electrical source. In this way, the local utility will still be responsible to service and repair the lines to the residence. Obviously, pricing will be important in terms of this choice, but deregulation will make this option possible.

"Green" power's cousin is "clean" power. "Clean," or "cleen" as it has been coined, power is generated from resources that are not considered part of the renewable power technology. "Cleen" power includes power generated from large hydroelectric sources and newer high-efficiency (low-emission) natural gas plants. While these power sources are not "green," they do have a lower impact on the environment than conventional power plants such as coal, oil, or nuclear. Renewable energy is produced from natural resources such as water, sun, and wind. These renewable energy sources are an almost limitless supply of power. Burning fossil fuels, a nonrenewable resource, to generate electricity results in air pollution and greenhouse gases. Air pollution is harmful to human health, and many experts say that greenhouse gases are leading to global warming. Renewable resources produce little or no pollution.

With consumers requesting "green" or "cleen" power, and with pricing being more competitive, the power source companies will have to listen. They will have to listen to these requests because if they do not, someone else will. Deregulation actually could spawn tremendous growth in this area, which ultimately will lead to a cleaner environment, slow global warming, and curtail the depletion of our natural resources and fossil fuels. As a species, humans have the brain power and ability to do remarkable things. Perhaps a beautiful by-product of deregulation can be the weaning of the world from fossil fuel use. Future generations may finally benefit. For years, the drawback to "green" power was the cost. With deregulation, costs will become more competitive. Remember, it could be worse—we could have an electrical power distribution system similar to that in Africa. Costs per kilowatthour approaching 50 cents are not uncommon, and the reliability of this expensive service is usually poor.

Basic Electricity

Since most home owners, and even many home builders, do not understand basic electricity, it is probably wise to gain a better understanding. In order to attempt to save energy with all those energy systems based on electricity, the home owner must know the basic concepts. The home builder usually subcontracts the electrical work of the project to an electrical specialist. By doing so, the home builder does not usually have all the answers and details. So both the home owner and the home builder need more information concerning residential electrical systems. In this manner, a better relationship between electrical voltage and current with respect to power can be established. Electrical circuits, lighting, appliances, and all other electrical components should be defined. A good, basic understanding of the generation, delivery, and use of electricity makes that monthly bill justifiable. Knowing what's going on within the residence with all that electricity also makes sense too.

Electricity

Electricity and magnetism are the basic elements behind the scenes in our homes. They were once thought to be two separate forces. However, Albert Einstein's theory of relativity showed that the two were very much interrelated. Electricity is produced from the movement of electrical charges, or electrons, whereas magnetism is created by those charges in motion and reacting with other elements. Electromagnetic forces and magnetic flux are all a result of the presence of electricity. Residential energy systems, in some way or another, rely on electricity.

With energy systems aside, most of what goes on in the home is a direct result of electricity. Electricity is a form of energy that consists of mutually attracted protons and electrons (positively and negatively charged particles). Therefore, in every home, electrons are flowing, making heat, light, and electric motors run. This electrical function may be apparent and, then again, may not be so obvious. Even communications is a form of low-level electricity traveling over conductors to pass data. In the case of the telephone, power is required to provide a dial tone and to transduce the voice signal to an electrical signal. Any motion, on the other hand (e.g., fans, pumps, refrigerator compressors, etc.), is caused by an electric prime mover or motor generated by levels of electrical energy. It all gets traced back to the power-producing utilities, and it all costs money—it is *not* free.

How Electrical Power Is Generated: An Overview

As can be seen in Fig. 9.4, electrical power is generated by a power plant. The primary methods of generating electricity are from coal-fired, oil-fired, and nuclear power plants. There are many other ways in which electricity is produced (hydroelectric, solar, etc.), but 80 to 85 percent of all electricity produced in this country is by fossil fuel or nuclear power plants. This is why

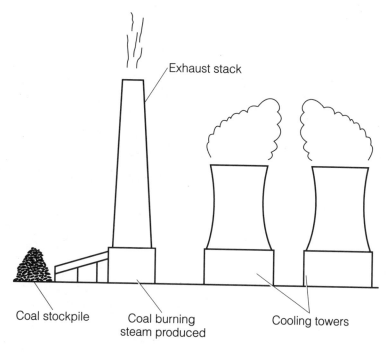

Exhaust stack

Coal stockpile

Coal burning
steam produced

Cooling towers

■ **9.4** *Power plant power generation.*

electricity should be conserved and used wisely. Until it can be produced without first using a valuable natural resource, it must not be wasted.

The electricity used in most residences is produced as alternating current (ac). It's production is illustrated in Fig. 9.5. As the turbine turns within the magnetic field, a voltage is produced at every point of a given revolution. The polarity changes during half the revolution, thus giving us alternating current. The magnitude of the voltage also changes, as can be seen in Fig. 9.6. Thus the usuable or *effective* power is calculated by the root mean square (RMS) of the peak. This is a trigonometric function, and the RMS value is generally 70.7 percent of the peak voltage.

In ac power transmission and distribution, the product of the RMS voltage and amperage is *apparent* power. When the applied voltage and the current are in phase with each other, the apparent power is equal to the effective power, i.e., the real power delivered to or consumed by the load. If the current lags or leads the applied voltage, the apparent power is greater than the effective power. Only effective power, i.e., the real power delivered to or consumed by the load, is expressed in watts. Apparent power is properly expressed only in voltamperes, never watts. Real power is measured in watts, so we are dealing with real power. Reactive power is in voltamperes reactive, and apparent power is in voltamperes. Thus the elec-

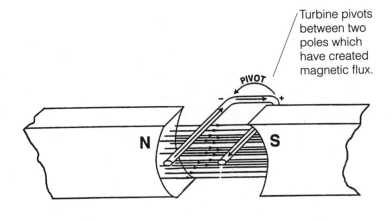

Turbine pivots
between two
poles which
have created
magnetic flux.

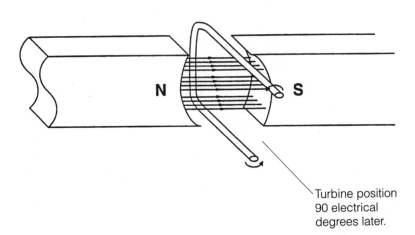

Turbine position
90 electrical
degrees later.

■ **9.5** *Alternating current turbine action.*

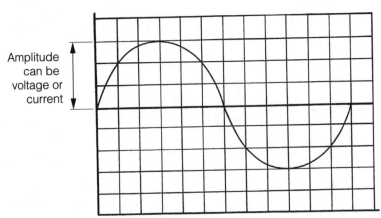

Amplitude
can be
voltage or
current

■ **9.6** *Amplitude.*

tricity delivered to your home is constantly changing polarity (positive or minus), and it is also changing voltage every cycle. From the power plant, over the power cables, and through large circuit breakers and transformers, electricity is brought to the residence—ready to be used and, unfortunately, a lot of the time wasted.

Once the power enters the home, it is distributed first through the electric company's meter and then to the fuse box or electrical panel (housing all the circuit breakers for the different circuits). These devices actually provide the usable voltages, 115 V, 220 V, single-phase being the most common. In addition, lower voltages still, such as 24 V and lower, are required for many appliances and most computerized systems. Most of the equipment commercially available in the United States requires one of the previously mentioned supply voltages. Three-phase ac power is commonplace within industrial and larger commercial facilities. This book will deal mainly with single-phase power. Let it be known that there will always be some piece of equipment or odd supply that is a voltage not previously mentioned. *The voltage value supplied to an appliance must match fairly closely the appliance's rating or else the appliance will not work or could be damaged!* This is where transformers come into play. Transformers are actually in use throughout the home. They are busy changing 115-V ac, single-phase electricity into 12-V dc power for the answering machine, telephone, and calculator. They plug into the wall outlet and run the device, giving off some losses as heat as the induced voltage is created. Transformer construction, use, and operation are discussed in further detail later in this chapter.

Electricity is either found in direct-current (dc) or alternating-current (ac) forms. There are electric motors and other pieces of equipment that can run from a dc source or an ac source. Many power supplies within the home are actually transformers. Frequently, these transformers are stepping the voltage that enters the house down to some lesser value and sometimes are converting it to direct current (dc). Many computers and microprocessor-driven appliances use low-voltage dc throughout their processor circuits. The home's electricity will be used more for these types of residential applications in the future.

Electrical Calculations

There are three elements in an electric circuit: voltage, amperage, and resistance. *Voltage*, usually shown as V or E, is the force that causes electrons to flow. *Amperage*, or *current* is expressed as I or A and is the actual flow of electrons, in which the unit of measure is amperes. *Resistance, R* or ohms, is the opposition to current flow. These three elements make Ohm's law, from which many basic electric circuit calculations can be made (Fig.

9.7). Ohm's law is adequate for dc circuit analysis and for some ac circuit analyses. Three-phase power, however, tends to be a bit more complicated. Figure 9.7 shows an *E, I, R* pie chart, which is useful when trying to remember the equations:

$$E = IR \quad \text{or} \quad \text{volts} = \text{amps} \times \text{ohms}$$
$$I = E/R \quad \text{or} \quad \text{amperes} = \text{volts/ohms}$$
$$R = E/I \quad \text{or} \quad \text{ohms} = \text{volts/amps}$$

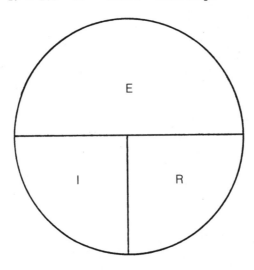

■ **9.7** *Ohms law pie chart,* E = IR.

When using the pie chart, the *E* value is shown above the *I* and *R* values. This represents *E* being divisible by either *I* or *R* when solving for either *I* or *R*. Likewise, to find *E*, it is necessary to multiply *I* times *R*, as is true of the actual equation. The pie chart is an effective and easy way to remember the aforementioned equations and their relationship to each other rather than memorizing the separate formulas. Ohm's law allows anyone to quickly determine whether or not there is an immediate discrepancy between voltage, resistance, or current values.

There are many derivations to Ohm's law, and there are other laws of electricity that must be understood. There are complete college degree programs that take the electrical engineer through all facets of circuit design and analysis. This makes for a plethora of calculations that can be performed involving higher levels of mathematics and calculus. Several analyses and transformations are used to prove and disprove certain issues. However, for our purposes, the basics and some rules of thumb are going to be the tools by which we, as home owners, can conserve electricity. Also, we shall look at certain analyses, postulates, and transformations only to indicate how and why they are important to the average person.

One such transformation is called the *Laplace transformation*. It is a mathematical integration used in harmonic analyses and other issues related to electricity, magnetics, and gravitational concerns. Another is the *Fourier analysis*, which is also useful in harmonic and waveform investigation, along with heat or energy transfer. The names for these analyses come from famous mathematicians, and many engineers and scholars will refer to these theorems. Another electrical law is *Kirchoff's law of current and voltage* for ac and dc circuits. While the intent of this chapter is not to mold the reader into an electrical engineer, many of these concepts and electrical laws come up more and more in normal discussions. This is due perhaps to the fact that so much more of our daily lives is affected by electronic and electrical equipment. Additionally, electricity is more than just voltage and current. It is the *power* that does work for us.

Electrical power P, or the rate of doing work, is measured in watts and is expressed in formula form for dc circuits as P (watts) $= V \times A$. We have to differentiate between equations for dc and equations for ac circuits mainly due to the steady-state conditions that exist with dc circuits and the non-steady-state conditions that exist with ac power. Ac circuits have to be analyzed differently when three phases of power are involved. Ac power has to be averaged and RMS (root mean square) values provided in order to get proper results (Fig. 9.8). An example of determining power for a light bulb is as follows.

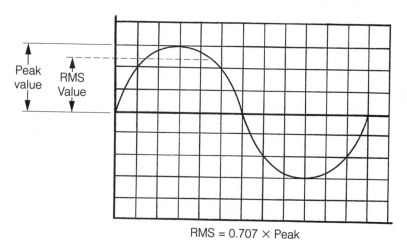

RMS = 0.707 × Peak

■ **9.8** *RMS current (root mean squared).*

Power Example: Since a light bulb is nothing more than a resistor, its current can be quickly determined using the $P = VA$ equation. If the light bulb is rated at 100 W, and the supply voltage is 120 V, then $I = 100$ W/120 V, or 0.83 amps.

Electrical power used to run a home's furnace blower, heat pump compressor, or even the washing machine is expressed in terms of horsepower. This is the size of the motor. Horsepower for a given dc circuit is equal to volts × amps × the efficiency /746. There are 746 W of power to 1 horsepower. The mode today is to have equipment as near to 100 percent efficiency as possible. This is factored into the equation

$$\text{efficiency} = \frac{746 \times \text{output horsepower}}{\text{input watts}}$$

which can be used for ac circuits and is a good indicator of where a particular process or piece of equipment is relative to its cost and productive output. Look at all the electrical equipment within your house and try to define the efficiency of those appliances driven by electric motors (e.g., are any of them warm? If so, then that heat energy is a loss to that part of the system).

Washing machine	Dryer
Furnace blower	Whole-house fan
Dishwasher	Heat-pump compressor
Refrigerator compressor	Freezer compressor
Fireplace fan	Water pump
Electric lawn mower	Electric weed whacker
Jacuzzi pump	Swimming pool filter/pump

Ironically, these are just a few of the electrical appliances that run a good portion of a day and use most of the home's electricity. Obviously, the list does not count lighting, electrical heating, and all the other appliances that might carry a small motor (fax machines, VCRs, printers, pencil sharpeners, etc.).

Frequency and Amplitude

Within the home today there are many components that operate on electricity of different frequencies. Most of the appliances are supplied and function on 60-Hz power, but many take the power conversion a step or two further. Computers use high-frequency switches of low-voltage electricity to perform all their tasks. This electricity is typically used in the direct-current (dc) state, while many other appliances use straight alternating-current (ac) power. Throughout the residence, electrical power is changed either in frequency, type (ac or dc), or strength (amplitude). As more homes are equipped with computers, microprocessor-based appliances, and automatic controls, home owners and home builders will need to understand the advantages and by-products of this trend.

The power that is supplied to the home is single phase and has some voltage rating. The other factor is the frequency. *Frequency* is the number of electrical pulses transmitted over a given period of time. Frequency is expressed in hertz, or its abbreviation Hz. For example, most power in the United States is in the 60-Hz range, whereas in Europe the frequency range is 50 Hz. Most homes are provided with 60-Hz power. This means that every second there are 60 pulses of electricity through a given point in a wire. Figure 9.9 shows a typical waveform depicting the frequency portion of the wave and the other necessary portion, called *amplitude*. This wave is actually in sinewave form. For every electrical wave, there must be a corresponding amplitude in order to provide any usable power. This amplitude is often in the form of voltage. This wave has two time-based components called *periods* or *cycles*. A *half period* is one of the halves of the wave and is sometimes called a *half-cycle*. This is illustrated in Fig. 9.9 and is a typical half-cycle for alternating current (ac). Figure 9.10 is an example of a square wave such as produced by a variable-frequency drive. Home electricity use will see more and more square waves and pulsing electronic equipment. Outages sometimes are expressed in cycles, or portions of a second. If the lights flicker, then most likely either the amplitude for a given cycle or so dropped or a complete cycle or wave was not even present. Not many people look at this situation in this sense. All they know is that the lights flickered, not really knowing why. If the condition persists, then a call to the electric company is in order.

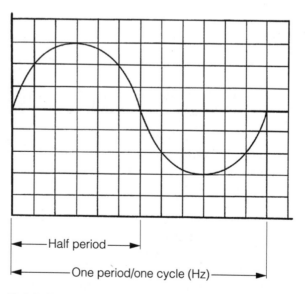

◼ **9.9** *Frequency.*

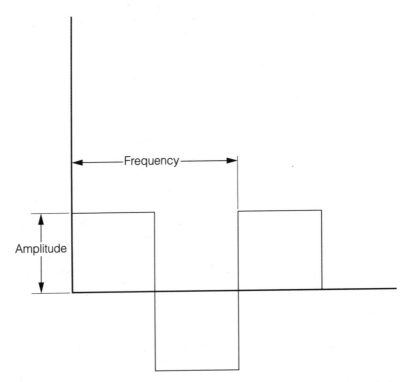

■ **9.10** *An example of square waves or pulses—in the house control circuitry.*

Electrical Hardware

Let us take a look at some of the components that might be found in the home today, especially in the common electrical circuit. These devices all perform a specific function and are commonly found throughout the home and its appliances without our usually giving them a second thought.

The most prevalent electrical component seen in the home is the resistor. Its rating is in ohms or R, which can take on many shapes and sizes. Since electricity is flowing through so many different devices these days, there is resistance in many areas of the home that we do not often consider. Lights, motors, and most other electrically driven components have some resistance value. On a smaller, physical scale, looking at the common printed circuit board (PCB), one will find possibly hundreds of small resistors. Lights are resistors and even the copper or aluminum wire carrying the electricity about the house is a resistor! Over time, some of these resistive values change with repeated heating and cooling of the components. A good example of this is the old radio amplifier/receiver. There is typically a good amount of heat coming from these devices, and over time the electrical usage will change to perform the same output—all because the component's resistance has changed. Because impeding the flow of electricity is sometimes desirable, resistors play a big role in making electrical and electronic devices perform as desired.

Likewise, resistance also can cause heat loss, which can result in both energy loss and a safety hazard. If unwanted resistance is present in a circuit, then the home owner may have to take measures to correct it. Several factors affect resistance, and Tables 9.1 and 9.2 offer the backup data for the following factors:

1. *Temperature.* An increase in temperature of any material causes greater vibration of the ions, thus making it more difficult for electrons to pass. Resistance is increased.

2. *Material type.* The more free electrons, the lower is the resistance of a material. Copper is a better conductor than aluminum.

3. *Length of wire or conductor.* The longer the material, the further electrons must travel in the resistance medium. Resistance is increased the longer a conductor is run.

4. *The thickness or diameter of the material.* The greater the thickness of a conductor, the more free electrons flow, thus offering less resistance to the flow of current.

For example, most wire carrying electrical power throughout the house is 12 or 14 gauge (AWG), which designates its diameter and thickness. This same wire thickness can have a different resistive value if it is copper than if it is aluminum. Many times, especially in older homes, the existing wiring is asked to carry more electric current. The wire may get warm or even hot. This is lost energy due to heat! This obviously is not safe either, since a fire could ensue. This condition should be corrected either by reducing the load on that circuit or replacing the wire with a larger diameter wire and better material.

Another source of lost energy in the conventional house circuit occurs wherever wire terminations are made. If a connection is not made firm and tight, then eddy current losses occur. The lugs, terminals, or nuts get very hot, and again, this is not a safe condition. Check these terminations (with the power off), and replace old fuse boxes, bad switches, and junction boxes wherever applicable. Have your home inspected from time to time by a qualified electrical inspector. He or she can recommend what needs to be done to make the residence more energy efficient and, more important, safe. Have qualified personnel do the work also.

■ **Table 9.1 Diameter, Resistance, and Weight of Copper Wire**

Wire Size/ AWG	Diameter/ Nominal inch	Resistance in Ohms per Thousand Feet	Weight	
			Lb per 1000 Ft	Ft per Lb
1	0.2893	0.1239	253.3	3.947
2	0.2576	0.1563	200.9	4.978
3	0.2294	0.1971	159.3	6.278
4	0.2043	0.2485	126.3	7.915
5	0.1819	0.3134	100.2	9.984
6	0.1620	0.3952	79.44	12.59
7	0.1443	0.4981	63.03	15.87
8	0.1285	0.6281	49.98	20.01
9	0.1144	0.7925	39.62	25.24
10	0.1019	0.9988	31.43	31.82
11	0.0907	1.26	24.9	40.2
12	0.0808	1.59	19.8	50.6
13	0.0720	2.00	15.7	63.7
14	0.0641	2.52	12.4	80.4
15	0.0571	3.18	9.87	101
16	0.0508	4.02	7.81	128
17	0.0453	5.05	6.21	161
18	0.0403	6.39	4.92	203
19	0.0359	8.05	3.90	256
20	0.0320	10.1	3.10	323
21	0.0285	12.8	2.46	407
22	0.0253	16.2	1.94	516
23	0.0226	20.3	1.55	647
24	0.0201	25.7	1.22	818
25	0.0179	32.4	0.970	1030
1/0	0.3249	0.09825	319.5	3.130
2/0	0.3648	0.07793	402.8	2.482
3/0	0.4096	0.06182	507.8	1.969
4/0	0.4600	0.04901	640.5	1.561

Parallel and Series Circuits

All the electrical wiring within the house is routed such that each circuit is parallel or series in structure. Many electrical devices, such as resistors, have to be strategically located in an electric circuit. These resistors can be the residence's lights or motors or other loads being electrically powered. It also should be noted that there are subtle differences between ac and dc circuits when utilizing series or parallel schemes. Resistor values in a circuit can be calculated as follows.

For example, Fig. 9.11 shows a series circuit. If $R = 6$ ohms and $R = 8$ ohms, then the total resistance in that circuit is $R + R = 14$ ohms. Conversely, in

■ Table 9.2 Diameter, Resistance and Weight of Aluminum Wire

Wire Size/ AWG	Diameter/ Nominal inch	Resistance in Ohms per Thousand Feet	Weight	
			Lb per 1000 Ft	Ft per Lb
1	0.2893	0.2005	77.02	12.98
2	0.2576	0.2529	61.08	16.37
3	0.2294	0.3189	48.44	20.65
4	0.2043	0.4021	38.40	26.04
5	0.1819	0.5072	30.47	32.82
6	0.1620	0.6395	24.15	41.41
7	0.1443	0.8060	19.16	52.19
8	0.1285	1.016	15.20	65.79
9	0.1144	1.282	12.05	82.99
10	0.1019	1.616	9.56	105
11	0.0907	2.04	7.57	132
12	0.0808	2.57	6.02	166
13	0.0720	3.24	4.77	210
14	0.0641	4.08	3.77	265
15	0.0571	5.15	3.00	333
16	0.0508	6.50	2.37	422
17	0.0453	8.18	1.89	529
18	0.0403	10.3	1.50	666
19	0.0359	13.0	1.19	840
20	0.0320	16.4	0.943	1060
21	0.0285	20.7	0.748	1340
22	0.0253	26.2	0.590	1690
23	0.0226	32.9	0.471	2120
24	0.0201	41.5	0.371	2700
25	0.0179	52.4	0.295	3390
1/0	0.3249	0.1590	97.14	10.29
2/0	0.3648	0.1261	122.5	8.163
3/0	0.4096	0.1000	154.4	6.478
4/0	0.4600	0.07930	194.7	5.135

the parallel circuit shown in Fig. 9.12, the same values for R and R exist. However, electrically, current flows differently (the paths of least resistance), and we calculate the total resistance accordingly:

$$R = \frac{(6)(8)}{6 + 8} = \frac{48}{14} = 3.43 \text{ ohms}$$

Thus it can be seen that in one scheme we can get 14 ohms of resistance and in another 3.43 ohms.

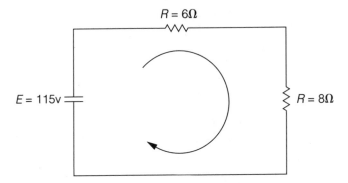

■ **9.11** *A series circuit.*

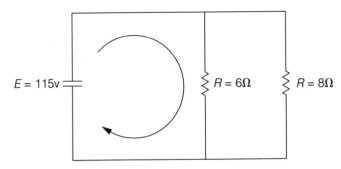

■ **9.12** *A parallel circuit.*

The best way for the home owner, who does not aspire to be an electrical technician, to understand the concept of series and parallel circuits is to think about Christmas tree lights. For years, lights were made such that if one light bulb burned out, the circuit was opened and all the lights went out. Today, most lights are manufactured as parallel circuits so that electricity will flow to the other bulbs even if one is out. Much of the home is wired in this fashion. It would be very inconvenient if a light bulb burnt out and all the house's appliances and other lights shut off!

Another electrical hardware component is the *capacitor*. This component is capable of storing electrical energy. It is sometimes referred to as a *condenser* because it consists of two conducting materials separated by a dielectric, or insulating, material. The conducting materials get charged, one positively and one negatively, thus creating a potential between them. The size and type of material of the conductors, the distance between, and the applied voltage determine the capacitance, in farads (one coulomb per volt), a coulomb being a basic unit of electrical charge. One farad is a large amount of capacitance. Usually, common capacitances are in the microfarad range. Uses of capacitors include placing them in circuits for tuning purposes and even as batteries for stored electrical power. Many televisions and other electronic devices in the home carry a capacitive charge to them because a capacitor is built into the appliance. This is why there may be a warning to not open the casing because there is an electric shock hazard.

When discussing electronics, it is necessary to include relays, diodes, and inductors because they all play an integral role in the automated home. Relays have been used for a very long time in controlling machines and processes. Relays work on the principle of electromagnetism. This principle allows the relay to function as a device that can either automate or control from a remote location any electric flow. There are basically two circuits, one the relay circuit and the other called the *energizing circuit.* These work together to open or close a switch. Enough energy is provided via the relay circuit to magnetize the appropriate element and thus complete the other circuit. This other circuit can be and usually is of a higher voltage and carries more current than the relay circuit. The operation of the common contactor can be likened to that of the conventional relay.

The relay is still very much an integral part of the home automation process today. Its switching is an important function in most processes, and years ago, much of that era's attempt at automation was by relay logic. Relay logic utilized many interlocking relays and contactors to control motors and processes. If a motor was to start, then permissive relays had to be in place and energized to allow that start. Today, more equipment is controlled by computers and microprocessors, which, in effect, do the switching in software.

Next, we shall look at diodes. A *diode* is a solid-state rectifier that has an anode, the positive electrode, and a cathode, the negative electrode. These nodes allow electricity to flow in one direction only. Diodes are used commonly to convert ac voltage to dc. The diode, in effect, acts as a valve for electricity. A diode is shown in Fig. 9.13 with its basic components and accepted symbol. This diode might be used for voltage regulation.

It is very probable that we have all gazed on a light-emitting diode (LED) at one time or another. These red or green lit displays are very common in

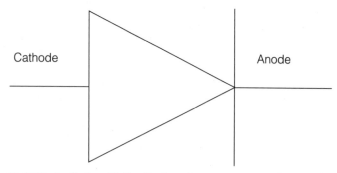

■ **9.13** *A diode with its discharging component, the anode, and the cathode.*

much of the electronic componentry in factories and even in our homes. They are made from materials that provide energy bands of different wavelengths for each color. LEDs are made from semiconductor materials such as gallium arsenide phosphide or gallium phosphide. These materials produce the red- or green-colored LEDs. There is a newer LED out that is blue. It is made from silicon carbide. When combined with red and green colors, the blue will complete the RGB package. This combination can produce white light. Additionally, varying current through these RGB LEDs will allow for many other colors. The possibilities are almost endless.

Another component often found in the home's electrical circuitry is the inductor. *Inductors* are devices used to control current and the associated magnetic fields relative to the currents for a given period of time. Inductance is measured in units called *henries*, which equate to one volt per one amp per second. Typically, an inductor consists of a coil of conducting material with a specific size and shape. This material is coiled around a core, most often of soft iron, and sometimes the inductor is called a *choke* for this reason. Most often the inductor is called a *transformer* or *reactor*. It is often used to slow the rate of rising current and for noise suppression.

Transformers

Transformers are an integral part of home and electronics. As a matter of fact, outside your house (somewhere on your street) is a transformer. It is supplying your home with the ac power necessary to power everything within your house. Without these specialized inductors, no usable electricity would be available at lower voltage levels. A transformer's simplified construction is shown in Fig. 9.14. Basically, transformers work on the principle of mutual inductance; hence they are classified as inductors. In order

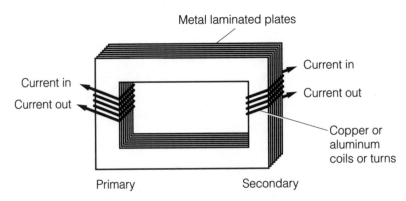

■ **9.14** *A common transformer with primary and secondary coils.*

for a transformer to work, there must be two coils, positioned such that the flux change that occurs in one of the coils induces a voltage across each coil. Typically, the coil that is connected to the electrical source is the *primary*, and the coil that is applied to the load is the *secondary*. Transformers are rated in voltamperes (VA) and kilovoltamperes (kVA). There are dry-type and oil-filled, with the size sometimes dictating which type to apply. There are also iron-core types, air-core types, and units that are called *autotransformers*.

Transformers can be used as isolation devices and step-up or step-down devices. Often it is desirable to include a transformer with similar primary and secondary voltages mainly to isolate a section of the electrical system from another. In this way, there is no direct physical connection, and thus conditions such as ground faults can be protected against traveling throughout an electrical system and destroying components along the way. Also, the step-down or step-up transformers used to match voltages in a given system also will offer isolation. Many of today's home appliances and electronic equipment requires 230-, 115-, 48-, 24-, 12-, or 9-V power to function properly. Transformers or small power supplies furnish us with these voltages. They also give off a good deal of heat and thus waste a good amount of energy. It may be worthwhile to count up all the small transformers within your home, such as all those devices on telephones, fax machines, answering machines, printers, and so on. Maybe future electrical systems for the residence will have a dedicated low-voltage line and do away with all the individual transformers. This may save a good deal of electrical energy.

Lighting

If we remove any type of electric heating, electric water heating, and any freezer-type equipment from the total energy load of a residence, the lighting load now tends to be one of the highest electrical loads within the home. While lighting represents approximately 6 to 7 percent of the total amount of kilowatthours used in a typical residence, concerns over which kinds of lighting are better and how to ensure that lights are off when rooms are unoccupied are still there. We have all instructed someone during the course of our life to "turn the light out when you're finished" so that valuable electricity can be conserved. It is definitely very important to only use lights whenever a room is occupied and when necessary, but the lighting load in a typical residence is not that great. Of course, this will vary from household to household (based on the number of occupants, size of the house, hours of daylight per year, etc.) and maybe in some cases sophisticated controls to turn lights off are in order. However, for the most part, the premise should be consistent with this book's theme—any and all energy savings are welcome; every little bit helps.

Fluorescent Versus Incandescent

This is no longer an argument: Fluorescent lighting is more economical from an energy-savings standpoint than is incandescent lighting. Compact fluorescent lights have revolutionized energy-efficient lighting. They only use 30 percent as much electricity as incandescent lights and last up to 10 times longer. They work like standard tube fluorescent lights using an electrode, a metal filament that emits electrons in the fluorescent lamp itself. The fluorescent tube is small and folded or bent over to concentrate the light. The compact design allows them to be used in place of incandescent light bulbs. A single 30-W compact fluorescent lamp used in place of a 100-W incandescent lamp saves about $55 over its lifetime. Compact fluorescents may cost more upfront to buy, but you save money in the long run because they use less than a third as much electricity and last almost 10 to 12 times longer than standard incandescent lamps. They should always be considered in fixtures that need to be "on" for long periods of time.

The fluorescent lamp is a gas-filled lamp in which light is produced by the interaction of an arc with phosphors lining the lamp's glass tube. The fluorescent light current is the path over which electric current flows to operate fluorescent lamps. Three major types of fluorescent lighting circuits are in use today: preheat, instant start (slimline), and rapid start. These electrodeless lamps actually were first developed by Nikola Tesla in 1891 but were not available commercially until recently. New developments in electronic technology paved the way for these systems, which use an induction coil powered by a radiofrequency power supply to create a magnetic field. This process excites the mercury in the lamp to create ultraviolet energy, which, in turn, excites a phosphor coating on the inside of the lamp to emit light.

However, incandescent lighting will remain with us for many years to come. Maybe in time fluorescent technology will be all that is available, or maybe a new technology will obsolete both. As for dimmers and variable-voltage supplies, these really do not save energy. These are commonly found on incandescent fixtures, and their use gives the impression that less electrical energy is being consumed. This may be the case at the bulb itself, but the lost energy is accumulating at the dimmer switch as heat. That is why a thin, aluminum heat sink frequently is mounted with the dimmer switch. The voltage is being throttled to the bulb. Thus the dimmer looks good and can make the mood right, but energy saving it is not.

Ballasts

A *ballast* is an electrical device used in fluorescent and high-intensity discharge (HID) fixtures. HID fixtures consist of the class of lights that use mercury vapor or low- and high-pressure sodium. These lights typically are used outdoors, whereas the fluorescent models typically are found inside. In either case, a ballast furnishes the necessary starting and operating current to the lamp for proper performance. Within the ballast there is a coil or windings of copper or aluminum wire surrounding a core. This is also an inductor, an electromagnet. There is also a metallic core, a component of electromagnetic ballast that is surrounded by the coil. The core is comprised of steel laminations or solid ferrite material. These laminations vibrate whenever the induced electromagnetic fields are of high strength. This is sometimes called *ballast hum*, and most of us have heard the sound. Ballasts are very efficient devices when compared with other lighting schemes. However, there are losses due to the vibrational noise and some heat when power is not converted to light energy.

An unwanted by-product of using ballasts is the harmonic distortion that is produced as these units do their job. Any devices that perform a power conversion tend to distort the electrical power being supplied. Electronic ballasts also operate at very high frequencies: 20,000 to 50,000 Hz. These are called *nonlinear loads*, and the more of them there are, the more dramatic the distortion becomes. Harmonic distortion is expressed in terms of the fundamental 60-Hz sine waveform. These harmonic waves travel along with the fundamental as unwanted frequencies or electrical noise. This phenomenon will continue to escalate as more nonlinear load equipment is installed within the home.

While the fluorescent light fixture and its electronic ballast bring an energy-efficient technology to the forefront, there is always some by-product that is undesirable. The extra up-front costs are higher than for incandescent lights, and the harmonic distortion could be a problem in the future. However, an incandescent lamp in which light is produced by a filament heated by an electric current is not practical from an energy standpoint. Look for fluorescent technology to improve and for costs to come down.

Off-Peak-Demand Systems

Many utilities offer a kilowatthour discount to their standard rates if power is used during nonpeak times of the day. Nonpeak hours typically are late evening through early morning when most consumers are not demanding electrical power. The period of high consumer demand is called the *peak*, and the electrical company sets its rates around these periods—and they are usually the highest then. However, some electric companies offer discounted prices for kilowatthours if the consumers use the power during the nonpeak or off-peak hours. Many commercial and industrial consumers take advantage of this routinely. Residential customers sometimes also can take advantage, as long as the electric company in their area offers a reduced rate (some do not). A telephone call to the utility will answer the question, since many electric companies do not advertise this rate.

To take advantage of off-peak or nonpeak rates, a home owner does not have to change his or her living schedule to wash and dry clothes at two in the morning (although this would yield the same results); rather, he or she can have passive heating systems in place to work whenever he or she is asleep. Also, automated home systems will allow the home owner to take advantage of these lower hourly rates, sometimes called *time-of-use pricing*. For example, if the rates are lower from 10:00 P.M. until 8:00 A.M., the consumer may want to install an electrical heating system that uses electricity during these hours (at the lower rate) to generate heat and have this heat stored in some "thermal mass" heat-storage system. Then, during the next day, after 8:00 A.M., the electricity can be off to the system, and heat can be released by natural radiation to heat the house. This is efficient use of energy, and home owners should look at their home's electrical usage, especially the time of day, and check to see if they could take advantage of lower rates. Each home will have its own individual applications.

Likewise, looking at the electric bill and correlating the appliances with the percentage of the bill, find out if there are other ways to use off-peak power. Electric hot water tanks can be placed on timers so that they only demand electricity late at night and early in the morning (before those showers). If the residence is equipped with a complete home automation system, then washing and drying clothes and running the dishwasher and other appliances also can be programmed to occur late at night. Of course, if the utility does not offer the lower rate, then these energy-saving approaches are futile. However, the utility does offer lower rates to industrial and some commercial customers. If the utility does this, then it probably is not allowed to discriminate, and therefore, you have a good case to request lower rates.

Technology and telecommunications are also allowing for better knowledge of electrical energy consumption with time-of-use data. New electric meters can be exchanged with the old meter that will transmit the time-of-use data automatically over the existing telephone line. This will not interfere with your existing phone service, security system, fax machine, computer equipment, or similar devices. These data then will be summarized in the monthly bill for analysis. These services can help pinpoint trends and reduce costs. The utility can alert the consumer to reduce electric bills with an incentive to shift use to off-peak hours or to basically try to conserve power. Additionally, the electric company can reduce operational costs and possibly reduce future requirements for new generation, transmission, and distribution facilities with this information.

The energy management concept called *time-of-use pricing* offers home owners an incentive to shift some of their electric consumption from a higher-priced on-peak period of the day to a lower-priced off-peak period. By shifting electric use to off-peak hours, an electric utility can optimize the use of its existing infrastructure and defer the need for future capital investments in generation, transmission, and distribution. With time-of-use pricing, electricity is charged based on the time during which it is used. Not all electric companies offer this, but it is certainly worth asking about. Peak, higher-priced periods are typically early afternoon until 8:00 or 9:00 P.M. during weekdays (the times when most individuals are up, home from work, and all needing to use electricity). Off-peak, lower-priced periods are typically all the remaining hours, with the real low-priced hours coming when most individuals are supposed to be asleep—midnight to morning and even 24 hours on Saturdays, Sundays, and holidays.

Variable-Frequency Drives in the Home

Have you ever stopped to count all the different ac motors running within your residence? Washing machines, dryers, water pumps, whole-house fans, furnace fans, spas, refrigerators, freezers, and heat pumps contain some of the larger motors in the home. Then there are the many small motors in other devices: dishwashers, fax machines, printers, swimming pool pumps, hair dryers, etc., and the list can go on. Not all these would be candidates for this new technology called *ac variable-frequency drives*, but many are. Today, we are on the verge of taking another step in technology in the home. Being able to run off single-phase power and with costs coming down dramatically, ac drives are coming home! These drives will control and protect a motor, save energy, and provide better performance for a good many of the things that move in our homes.

During the 1970s, the oil embargo changed the lives of all Americans. It actually changed the entire world. The world's energy economy was "kicked" into high gear, and so too was technology. We were in the midst of an energy crisis. Gasoline had to be rationed. Oil-related products saw prices go out of sight—virtually overnight. Something had to be done. No longer could we waste energy at the rates we were used to. Energy was to be conserved in all forms; automobiles, heating, and electrical equipment were the top three. Efficiencies had to increase on electrical equipment, and new ways of saving energy had to quickly be developed.

One technology that was available but seldom used was that of converting ac energy into dc and then inverting that energy back into ac and varying the frequency to an electric motor. This technology saved energy, but the cost to implement such a device was extremely high. Thus it sat in the shadows until the electric utility companies nearly panicked. Existing oil-burning power plants had to be converted to coal, and new plants had to be coal- or nuclear-powered. No more oil-burning plants! And the price of electricity went up and kept going up. First, factories, schools, and hospitals had to react. Now it is time for the residential homeowners of electrical equipment to be on the alert.

What should be done in the residence to reduce energy consumption. One possibility was to make every piece of equipment, every appliance, as efficient as possible. Another was to install premium-efficiency ac motors whenever practical. Suppliers and manufacturers of heating and cooling equipment have recognized that they have to do this. But what could be done with all the ac motors existing and running in the home, especially those motors which always run full speed but incorporate mechanical means of slowing or reducing the flows of liquids or gases? These were the motors running all the fans and pumps, constantly running at full speed whether they had to or not. These were good motors, too, not justifying being replaced by an energy-efficient motor because that would not solve the main problem. The problem was that many hours per day these existing motors did not have to run at full speed because the loads and demands were not that severe. In fact, over the period of a year, the motor's energy costs were exorbitant. Fan systems had mechanical dampers in place that could be adjusted to restrict the flow of air as necessary to get the desired flow—all the while as the motor was running full speed.

The same was true with the centrifugal pumps. The traditional method of getting the desired reduced flow was to close a valve somewhere in the pipe system. The motor still ran at full speed! In fact, these mechanical methods of reducing flow actually were hard on the system. The analogy is that of constantly fighting the full power of the electric motor. Dampers and valves eventually would fail and wear out, needing to be replaced. Weak locations in the piping system and the ductwork often ruptured, causing more maintenance. All this was the traditional method of installing and maintaining a fan or pump process. There was a better way, and the energy crisis made its implementation cost-effective and practical. Since then, the technology of ac drives has never looked back. As a matter of fact, ac drives have come down so much in cost that they are more commonplace than ever; using them within the residence will be the next huge market for these products.

The use of ac drives has proliferated not only because of their energy conservation but also because of their ease of incorporation into home automation control, their soft starting capability, and the ability to protect the motor. With this proliferation has come major strides in cost reduction (isn't competition nice?), packaging, and most of all, the features and technology of the drives themselves. Different types of drives have emerged, all useful in some application. Interestingly, ac drives got their main application thrust from industry and now have become commonplace in the commercial marketplace. As costs come down, they will even become prolific in the residential arena. This section seeks to define what these ac drives are, the different types, how to apply them, how to troubleshoot and repair them, and how to "speak the language of drives."

The word *drive* can mean different things to different people. We used to take a "drive" on a Sunday afternoon (before the energy crisis). Many interpret it to mean the device in your computer on which you store a program or place a floppy disk. Others may consider a drive to be all the mechanisms required to move part of a machine. Some may call this the *drive train*. Even when referring to it as an ac drive, many still do not have a clear understanding of what it is. The drive train is being powered by an ac motor, and one simply assumes that this is the "ac drive." Thus the clarifier should be that when referring to a drive that electrically changes the electrical input to a motor, it should be called an *electronic drive*. Better yet, calling the device an *electronic ac drive* is a further clarifier.

The electronic ac drive has many names to different people: *motor drive, variable-frequency drive* (VFD), *variable-speed drive* (VSD), *adjustable-speed drive* (ASD), *adjustable-frequency drive* (AFD), *volts-per-hertz drive, frequency drive*, or in many instances simply an *inverter* (a misnomer as we will discuss later). The terms *freq drive, VF drive*, and *varidrive* are still other terms that have been coined.

Electronic AC Drive Basics

The main components of an electronic ac drive are the power bridges and the control section. Figure 9.15 shows in simplified block form the two main power sections, the dc link, and the control scheme. The power bridges, the way the drive derives electrical feedback from the motor, and the drive's output waveform all describe the type of drive being used. Like their dc counterpart, all ac drives have to have a power section that converts ac power into dc power. This is called the *converter bridge* and is seen in Fig. 9.16. Sometimes called the "front end" of the ac drive, the converter is commonly a three-phase full-wave diode bridge. Compared with

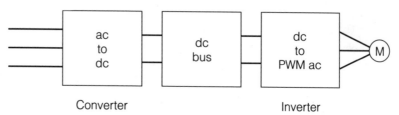

Converter   Inverter

■ **9.15** *Ac drive components.*

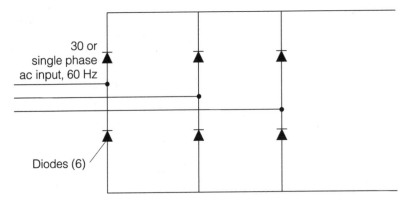

■ **9.16** *The converter section of a variable-frequency drive.*

the phase-controlled converters of older style ac drives, today's converter provides for improved power factor, better harmonic distortion back to the mains, and less sensitivity to the incoming phase sequencing.

The next component is the *dc bus* or *filter*. This section is common to all electronic ac drives in some form. This is the section of the drive's circuitry where many drive manufacturers filter the dc bus voltage. Capacitors or chokes are used to ensure that the desired dc voltage or dc current is supplied to the inverter section. Here at the dc bus, valuable protective functions take place. The dc voltage is monitored for surges and compared with a maximum limit so as to protect devices from overvoltages. Also, the dc bus can provide the "quick outlet" for braking energy to a bank of resistors whenever a motor becomes a generator. Often the dc bus is called *the intermediate circuit, bus,* or *link.*

The main portion of the electronic ac drive is the *inverter section* (Fig. 9.17). This power bridge actually is the differentiating component in drives from a basic standpoint. This is where that constant-voltage dc energy is inverted back to ac energy through the power semiconductor network. Dc drives do not have an inverter section, so by default, ac drives are more complex and expensive right away. However, having an inverter section is a good component to have. Without it, we cannot vary the frequency of an ac motor.

The inverter takes on many shapes and sizes. Its design is so important to the ac drive that many individuals have simply started calling ac drives inverters. This is not a proper name because the drive still has other functions going on such as converting ac to dc, filtering the dc, and controlling all these functions in addition to inverting dc to ac. The inverter section is also where most of the drive differences will lie between manufacturers' designs. Some designs use thyristors, whereas most modern inverters use some type of transistor. However, the inverter's principle of operation remains the same—change dc energy into ac.

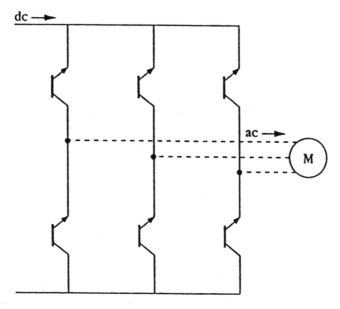

■ 9.17 *The inverter section of a variable-frequency drive.*

Inverters will be classified as voltage sourcing, current sourcing, or variable-voltage types. This has to do with the form of dc that the inverter receives from the dc bus. It also is a function of how the drive has been designed to "correct" its own electrical feedback loop. This loop is actually part of a comparison with inverter output to the motor and the motor's load. In order to continue driving the motor at the desired speed, the drive constantly must be correcting the motor's flux.

Early drive manufacturers used varying designs of inverters: thyristor-type semiconductors, different methods of cooling, different style heat sinks, and so on. The basic concept of the inverter, however, was the same, and eventually the thyristor gave way to the faster-switching transistor. Because the thyristor has to wait for the current passing through it to reach zero, it has become virtually obsolete in the design of inverter sections. The transistor, able to change from conductive to nonconductive almost instantaneously, has become the device of choice. Switching frequencies (the bread and butter of a frequency drive) now are in the 15-kHz range out to the motor.

Electronic AC Drive Types

Electronic ac drives end up being named and classified by their use, by their dc bus voltage or current sourcing, by their waveform (PWM or PAM), or by the type of power device used in their inverter section. By classifying an ac drive by its supply of voltage or current to the inverter, we get variable-voltage inverters (VVIs), voltage sourcing inverters (VSIs), and current sourcing inverters (CSIs). Ac drives can be called pulse-width-modulated (PWM) or pulse-amplitude-modulated (PAM) because the names describe the drive's output waveform. Lastly, we have ac drives being referred to as *transistorized, insulated-gate bipolar transistor* (IGBT) type, or even six-step SCR types, which describe the inverter devices being implemented. Electronic ac drives are usually classified by their output and the shape of that output's waveform. The main objective of the ac drive is to vary the speed of the motor while providing the closest approximation to a sine wave for current. After all, when an ac motor runs directly off 60-Hz power, the signal to the motor is a sine wave (as clean as the local utility can provide). Put a variable-speed drive in the circuit, and vary the frequency to get the desired speed. The energy savings are in the drive's ability to vary the frequency as it sends constant voltage pulses to the motor to operate and control it. This sounds simple enough, but the industry is continually striving to address all the side effects and to provide a pure system.

The pulse-width-modulated ac drive is the most common today. Most often it will integrate transistors into the inverter section to accomplish the switching pattern. The switching pattern is designed to control the width of the pulses out to the motor. The voltage waveform is shown in Fig. 9.18. The output frequency of a PWM drive is controlled by applying positive pulses in one half-period and negative pulses in the next half-period. The dc voltage is provided by an uncontrolled diode rectifier. Thus, by switching the inverter transistor devices on and off many times per half-cycle, a psuedosinusoidal current waveform is approximated. As was seen in Fig. 9.18, a six-pulse PWM inverter produces some harmonic content. As discussed earlier with electronic ballasts, there is some harmonic distortion associated with these drive motor controllers. These are, again, nonlinear

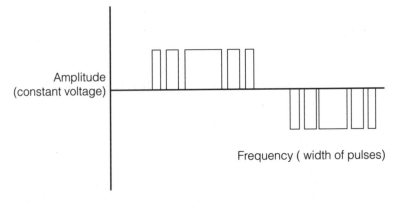

Amplitude (constant voltage)

Frequency (width of pulses)

■ **9.18** *Waveform for a variable-frequency ac drive.*

loads to the house electrical system. As they are implemented within the residence, harmonic distortion can occur.

As discussed earlier, the electronic ac drive is made up of a converter section, dc bus link, and an inverter section. Therefore, there are many converter designs, and the same is true for the inverter section. Generally speaking, the converter section, or the front end of the variable frequency ac drive, is the dc drive for a dc motor, with some modifications and a field regulator. The more common designs for the converter use diodes or thyristors for rectifying the ac incoming voltage into dc voltage. One is called the *diode rectifier*. In Fig. 9.19, the three-phase or single-phase incoming 60-Hz power is channeled into three legs of the converter circuit, each leg with two diodes. *Residentially, single-phase power is the norm, and the variable-frequency drive does not care if this is its input as long as it gets enough power to create a constant dc voltage.* From here

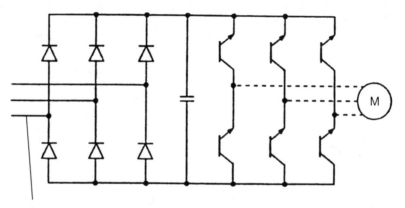

Three phase input

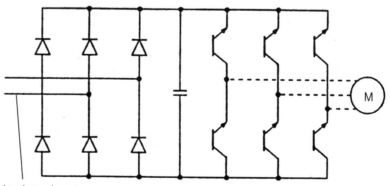

Single phase input

■ **9.19** *Three-phase or single-phase electric power into a variable-frequency drive.*

this constant dc voltage, through the dc link, goes to the inverter circuit to be changed into variable frequency ac to the motor. The diode rectifier is the most popular design because it is simple and inexpensive. Advantages of this type of rectifier include unity power factor, less distortion backfed to the supply, and resilience to noise in the converter itself. The biggest drawback to using diodes is the fact that no voltage can return to the source, therefore allowing no regeneration. Another separate power bridge must be added to the converter section if regeneration is required, and this can be expensive.

The heart of the variable-frequency drive is its inverter section. It is the most complex portion of the ac drive and hence is probably the reason that ac variable-frequency drives are sometimes called just *inverters*. There are many designs, and there has been much discussion on which is best. The answer is that they all are when selected for the right application. Some designs inherently will cost more than others, whereas other designs may have horsepower limitations or address harmonic distortion. Basically, most lower-horsepower inverters incorporate high-frequency transistors, whereas some higher-horsepower inverters do not. Individual power device costs and the paralleling of devices to get higher currents can be expensive.

The latest technological design method for inverters is based on the insulated-gate bipolar transistor (IGBT). This transistor has a combination of features provided by the MOS-FET transistor and the bipolar transistor. It has good current conductance with low losses. It possesses very high switching frequency and is easy to control. All in all, electronic ac drives simply have to function in this manner: Merely take three-phase or single-phase ac line frequency power, convert it to dc, and invert it back to variable-frequency ac. Sounds fairly straightforward and simple, right? Not so, as we have seen. Higher voltages and currents, transients, misapplications, bad motors, and so on make for interesting problems and challenges. However, the technology does keep evolving, and it is also getting better. Time will tell!

How Do Electronic AC Drives Work?

Understanding how the variable-frequency drive converts and inverts energy is valuable, but this is only part of the equation. Not only does the electronic drive control the ac motor, but it also becomes part of the motor's electric circuit. The drive needs the motor, and the motor needs the drive. The proper operation of the motor-drive system is critical to each supplying the other with voltage and/or current, and vice versa.

In order to really understand the application and operation of an electronic ac drive, one must first understand how an ac motor functions. To a drive there are motor basics that have to be understood (and furnished to the drive at setup). Motor-speed concepts and speed-versus-torque concepts are critical. Consider the following formula for an ac motor:

$$\text{Synchronous speed} = \frac{120 \times \text{frequency}}{\text{number of poles}}$$

Here, the value of 120 is a constant and cannot be changed. It is derived from the electrical relationship given to synchronous machines with fixed poles, location of those poles in a given half-cycle, and the frequency (cycles per second and seconds per minute). Thus, for a given motor, the number of poles has to be constant; they are physically in place on the motor's rotor. Therefore, in order to change the speed, all that can be changed in the formula is the frequency. Which is exactly what a variable-frequency ac drive does! In the equation, if the frequency is 60 for normal 60-Hz supplied power, then the motor speed will be the maximum for an equating number of poles. This is referred to as the *synchronous speed*. As an example, 60 times 120 is 7200, and this goes in the numerator. A two-pole motor has just that—two magnetic poles, north and south. Therefore, 7200 divided by 2 equals 3600, and thus this is the speed, in revolutions per minute, that the motor will run at if it is applied 60-Hz power. Likewise, if the frequency is 30 Hz, then the resulting speed is half, or 1800 rpm. Table 9.3 shows synchronous speeds for various motors with different pole configurations. Those shown represent the most common configurations.

■ Table 9.3 Ac Motor Speed Chart

Number of Poles	Synchronous Speed (rpm)
2	3600
4	1800
6	1200
8	900
10	720
12	600

NOTE: To determine synchronous speed, use the formula: speed = 120 × frequency/number of poles.

The number of poles in the ac motor is fixed. These magnetic regions on the motor's rotating element, the rotor, have a permanent polarity, plus or minus. As the stator windings receive electrical energy from the ac drive, a change in magnetism takes place, and thus motion of the rotor takes place. How often this current flows through the windings is the frequency. It should be noted that when an ac drive first applies power to the ac motor, a certain amount of current is needed to get magnetic flux built up enough to even begin motion. This is called *magnetizing current* and must be accounted for by the drive before any torque-producing current can be used.

When given the task of driving an actual load, the actual shaft speed of an ac squirrel cage motor will be slower than the synchronous speed. This is called the *full-load speed* and is a function of a motor characteristic called *slip* and is typical of all ac induction motors. Because the motor is always dynamically correcting to maintain speed, when loaded, it lags behind in actual motor revolutions per minute. The percentage of slip can be found by using the following formula:

$$\text{slip percentage} = \frac{\text{synchronous motor rpm} - \text{full-load rpm}}{\text{synchronous motor rpm}} \times 100$$

The amount of slip in the typical NEMA B design squirrel cage motor is usually 2 to 3 percent of the motor synchronous speed. For example, an 1800 rpm synchronous-speed motor, when supplied three-phase power at 60 Hz and with no loading, will run at 1800 actual revolutions per minute. However, when the motor shaft is applied to a load, the actual speed will now be 1746 revolutions per minute. The speed at which a motor can run fully loaded is called *base speed*. A motor can run above or below its base speed by increasing or decreasing the frequency. As for loading and slip, these items must be factored into the application whenever running dramatically higher or lower than base speed. Also, NEMA A, C, and D motors carry different values of slip and therefore will have varying needs for current, particularly magnetizing current. Their slip percentage can range from 5 to 8 percent. This must be a consideration when applying the variable-frequency drive.

The motor slip is going to relate directly to the ability to drive any given load at a given speed. Torque can be said to equate to load, which equates to current. Therefore, there are given relationships to speed and torque. A typical speed/torque curve for a NEMA B design motor is shown in Fig. 9.20. Also depicted in Fig. 9.20 is a typical curve for any centrifugal or variable-torque load—one that might be found on a furnace fan or water pump within a home. Slip will remain constant anywhere on the curve as the frequency is reduced to achieve the desired speed. Slip is a critical element in controlling an ac motor, especially at low speeds. In essence, controlling slip means that the motor is under control. Electronic ac drives with volts-per-hertz capability can control slip very well down to low speeds. The biggest factor is the loading. Light or even centrifugal loads, such as furnace fans and even pumps, are much easier to control at slow speeds.

Torque is produced as the induction motor generates flux in its rotating field. This flux must remain constant to produce full-load torque. This is most important when running a motor at less than full speed. And since ac drives are used to provide slower running speeds, there must be a means of maintaining a constant flux in the air gap of the motor. This method of flux control is called the *volts-per-hertz ratio*. When changing the frequency

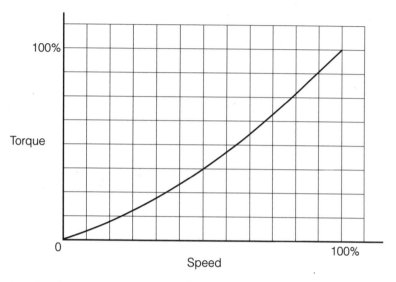

■ **9.20** *Speed versus torque curve.*

for speed control, so too must the voltage change, proportionally, to maintain good torque production at the motor. The volts-per-hertz ratio nominally is 7.6:1 for a 460-V, 60-Hz system (460/60 = 7.6). Thus, at half speed on a 460-V supplied system, the frequency is 30 Hz, and the corresponding voltage is 230 V to the motor from the drive. This pattern makes the ac drive save energy to the motor, but it is also very critical to performance. The variable-frequency drive tries to maintain this ratio because if the ratio increases or decreases as motor speed changes, then motor current can become unstable and torque can diminish. This is the reason that variable-frequency drives start to have control troubles below 10 Hz. Another method of increasing the voltage at low speeds to produce adequate torque is by incorporating a voltage-boost function available on most drives. However, if the motor is lightly loaded and voltage boost is enabled at low speeds, then an unstable, growling motor may be the result. Voltage boost should be used when loading is high and the motor must run at low speeds or start with a high load.

Our electronic ac variable-frequency drive also allows for motor operation into an extended speed range, sometimes called *overspeeding* or *overfrequency operation*. Sometimes the application requires that the motor to run beyond 60 Hz. Frequencies of 120 (twice base speed), 300, or 600 Hz and even beyond are possible with faster-switching inverters. Higher speeds can be achieved, but torque diminishes rapidly as the speed goes higher. The volts-per-hertz curve for this type of extended-speed operation can be seen in Fig. 9.21. Trying to maintain a constant ratio between the voltage and the frequency from zero to full speed is desirable in these kinds of applications.

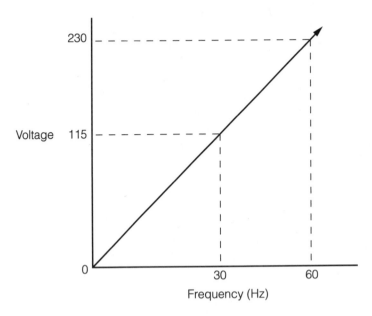

■ **9.21** *Linear volts per hertz curve.*

Where this capability really is a benefit is in pumping applications where getting a little more speed out of the motor can increase the flow just enough to satisfy the demand. When there is capacity in the motor and the ac drive can be programmed to do this, a new, larger motor does not have to be purchased and installed. The old motor does not have to be taken out. Often the ac drive can run 10 to 20 percent higher (66 to 72 Hz) in speed and make up for lost capacity in a flow and demand type of system. There is presently a need for this type of operation in many existing refrigeration systems that use old-style refrigerants—those being outlawed by regulatory agencies. These chlorofluorocarbon-containing gases are compressed by a motor. The new refrigerants, those allowed in place of the old, have different physical properties, and the motor has to speed up in order to get the desired flow and net capacity of the refrigeration system. Electronic ac drives may come to the rescue in many of these instances.

Energy Savings with Electronic AC Drives

With so many appliances and pieces of equipment within the home being motor driven, the ac variable-frequency drive will become commonplace over the next decade. While the size of most of the motors in a home is under 5 hp, the energy savings of all the small ac motors represent a good percentage of the home owner's monthly electric bill. In addition, the use of variable-frequency drive will allow for other functions and operations that previously were cost prohibitive. Fans or pumps can run for longer periods of time now that the reduction in speed comes through only using the energy required to run the motor at that speed. Before drives, dampers, valves, and other mechanical restrictors had to be used, and the motor ran full-out, using full power.

The reality of the energy savings attributed to applying electronic ac drives is evidenced by the rebate programs in place around North America. Many electric utility companies offer substantial rebates to end users in hopes that ac drives will be considered more often as an energy-saving alternative. The mere fact that the electric utilities are offering monetary incentives to implement ac drives is a sound testimonial to the fact that these devices actually save energy. And the results are usually outstanding! Of course, the utilities hope that the end user will be so happy with the energy savings that subsequent ac drives will be funded elsewhere.

The rebates are offered regional and periodically. They can range anywhere from $20 per horsepower to completely covering the purchase and installation of a drive system. Electric utilities have to offer these rebates. The biggest reason is because the prospects of bringing a new power plant online will take years of lobbying (once the Environmental Protection Agency, DER (Department of Environmental Resources), and other environmental groups give their approval) with the locality for the site; then add on the years of construction and phase-in and, of course, the costs—millions if not billions of dollars to get the new electricity down the wire. The better scenario is to entice home owners to conserve. Heating, ventilating, and air-conditioning markets as well as pump applications are perfect candidates for energy-saving electronic ac drives and even premium efficiency motors.

So why all the fuss over ac drives? Not only can they provide exceptional cost savings while operating, but they also can offer a power factor improvement, which further makes the utility happy. Ac drives limit inrush current to motors. They also provide a buffer, or filter, to the utility's supply such that line fluctuations caused by leading and lagging elements on the supply line do not appear as often and are not as severe. There will be no more penalties from the utility to the user of ac drives. This alone has caused companies to justify the cost of installing an ac drive. However, the real savings is in the actual operating costs saved over time. These costs are more dramatic in areas where electrical costs are very high.

An electronic ac drive can be used in place of several mechanical restrictors in most centrifugal fan or pump systems. With the fan system, an ac motor traditionally will run at full speed all the time, and the only way to slow the airflow is to open or close dampers or use inlet guide vanes. This was the method before drives and, surprisingly, is still a common method in many commercial buildings today. As for pump systems, the same reduction in flow was achieved by mechanical means. A pump motor will run at its full load speed. Discharge or bypass valves are placed in the piping system to be opened and closed, either manually most of the time or automatically when this mode can be afforded by the homeowner, to reduce the flow of liquid. These dampers and valves can get stuck, wear out, or corrode. Besides being maintenance items, they help in no way to save energy. That is where the ac drive comes in.

The ac drive is installed ahead of the ac motor in the electrical system. It now reduces the flow electronically. It slows the ac motor down and puts out only the voltage and current required to drive the load. And with centrifugal applications, the load is dramatically lowered as speed is lowered. These fans and pumps follow certain principles of physics called the *affinity laws*. These can be seen in Fig. 9.22 along with a curve showing how the ac drive changes output for various speeds to a NEMA B design motor. This curve shows that the ac drive can supply full voltage and current if required, but a centrifugal load does not require 100 percent or more torque except usually at full speed.

Affinity Laws
for
Centrifugal Fans and Pumps

$$Q2/Q1 = N2N1$$
Flow is proportional to speed

$$P2/P1 = (N2/N1)^2$$
Pressure (torque) is proportional to speed^2

$$HP2/HP1 = (N2/N1)^3$$
Power is proportional to speed^3

■ **9.22** *The affinity laws.*

The affinity laws, along with other basic formulas for fans and pumps, should be understood by the home owner, manufacturer of the equipment, or installer of the drive. This knowledge is useful in getting the full capability out of the variable-frequency drive. Even before applying the drive, it is sometimes necessary to completely justify the drive from an energy-savings viewpoint. An example of calculated savings is shown in Fig. 9.23. The minimal data required are the cost per kilowatthour, the motor size, the hours of operation, and the present means of flow restriction. Once these data are entered into the formula, the calculation is fairly straightforward. Sometimes the exercise is not even necessary if the electric company where the drive is being considered has a very high cost per kilowatthour. These are regions

Energy savings example;

Centrifugal water pump using a variable frequency drive cost savings example:

Givens:

The pump runs 12 hours per day, 7 days/week, 52 weeks/year (4368 hours total).

Motor is 2 Hp (Hp times 0.746 = KW, thus 2 × 0.746 = 1.492 KW).

Assume electricity cost is 9 cents per KW-Hr.

The pump flow is presently controlled by a discharge valve.

A VFD will save approximately 50% in KW operating the pump in a centrifugal manner.

Therefore:

(1.492 times 0.5) × (4368 Hrs.) × (0.09 cents/KW-Hr) = $438 savings/year.

Note:

A 2 HP variable frequency drive can be purchased for under $400. Therefore, a VFD will pay for itself during the first year.

■ **9.23** *Calculated savings.*

where there is high power consumption and limited production capacity. Thus the cost for electricity is high, and rebates probably are available.

For those home owners who enjoy technical data and mathematical challenges, the next section will provide just that. Other formulas for calculating the brake horsepower (bhp) are shown below. These formulas are necessary for calculating the horsepower in a new installation for fans or pumps. Otherwise, in existing installations, motor nameplate data are all that is needed. However, checking the existing brake horsepower to a recalculated brake horsepower will verify that the motor is not oversized (another waste of energy). The formulas are as follows:

Fan application formulas:

brake horsepower (bhp) =

$$\frac{\text{cubic feet per minute (cfm)} \times \text{pounds per square foot (lb/ft}^2)}{33,000 \times \text{fan efficiency}}$$

or when pounds per square inch (lb/in^2) are known,

brake horsepower (bhp)=
$$\frac{\text{cubic feet per minute (cfm)} \times \text{pounds per square inch (lb/in}^2)}{229 \times \text{fan efficiency}}$$

Pump application formulas

Brake horsepower (bhp) =
$$\frac{\text{gallons per minute (gpm)} \times \text{feet} \times \text{specific gravity}}{3957 \times \text{pump efficiency}}$$

or when pounds per square inch (lb/in^2) are known,

Brake horsepower (bhp) =

$$\frac{\text{gallons per minute (gpm)} \times \text{pounds per square inch (lb/in}^2) \times \text{specific gravity}}{1713 \times \text{pump efficiency}}$$

Where head in feet is equal to 2.31 pounds per square inch gauge (psig) and the specific gravity of water is equal to 1.0. Various specific gravities for other common liquids are shown in Table 9.4.

■ **Table 9.4 Various Specific Gravities for Common Liquids (English System of Units at 14.7 psia and 77°F)**

Liquid	Specific Gravity
Acetone	0.787
Alcohol, ethyl	0.787
Alcohol, methyl	0.789
Alcohol, propyl	0.802
Ammonia	0.826
Benzene	0.876
Carbon tetrachloride	1.590
Castor oil	0.960
Ethylene glycol	1.100
Fuel oil, heavy	0.906
Fuel oil, medium	0.852
Gasoline	0.721
Glycerine	1.263
Kerosene	0.823
Linseed oil	0.930
Mercury	13.60
Propane	0.495
Sea water	1.030
Turpentine	0.870
Water	1.000

Thus, if we were to predict what the energy savings would be for a motor whose brake horsepower was determined to be 5 bhp, the first step would be to convert that horsepower value into real power, or watts. Thus 5 times 0.746 (746 W equals 1 hp) equals 3.73 kW. A variable-frequency drive is assumed to be at least twice as efficient as a pump with a discharge valve, and therefore, we apply a ratio to each. Pump systems notoriously operate at 65 to 75 percent of their maximum flow rates, and by using pump curves, which can be supplied by any pump manufacturer, we can establish typical efficiencies and other data useful to confirm our ratios. Next, we calculate the power consumed by the pump using the discharge valve, and at a ratio of 0.98 to the drive at 0.5, we come up with 3.73 kW × 0.98 = 3.655 kW and 3.73 kW × 0.5 = 1.865 kW, respectively. Subtracting the smaller value from the larger, we now have the difference in energy between the two types of pump motor control. That difference is 3.655 − 1.865 = 1.79 kW. This 1.79 kW is the actual power that will be saved every hour. By applying two more factors, time and money (actual hours of operation and cost per kilowatthour), we can put a dollar value on the savings. If the pump runs 8 h/day, 5 days/week, then in 1 year it runs 2080 hours. If the utility charges 9 cents per kilowatthour, then we simply take 0.09 times 2080 times 1.79 kW and arrive at a savings of $335 per year. Considering that a 5-hp drive may cost $500 (using the rule of thumb of $100 per horsepower cost estimate), then saving $335 each year means that the drive will pay for itself in 18 months.

Obviously, the preceding example makes some assumptions. However, if the fan or pump runs 24 h/day and/or the electrical costs are higher, then the actual savings will be much greater. Likewise, if the motor is much higher in horsepower rating, the savings come back quicker too. The bottom line is that a variable-frequency drive will save money in a centrifugal application. How much and when the investment in the drive will be returned are answered in each individual application by hours of actual motor running, the cost per kilowatthour, and how good a deal one gets on that drive purchase. When a rebate is a possibility, then implementing the ac drive should be done as fast as possible. This is done for two important reasons: First, the rebate period will expire, and second, for every hour the motor runs without an ac drive controlling, more money is being wasted.

AC Drive Selection

As drives become more prevalent within the residential industry and markets, home owners will have to become more knowledgeable about actual drive products. They will have to know basic differences between types, know how to troubleshoot and service them to a degree, and maybe even how to program them. We all know that with every advancement in technology comes opportunity for service, training, and repair entities. These costs are going to be borne by the home owner, and thus many will undertake the challenge of understanding the whole breadth of the ac variable-speed drive product. These are the do-it-yourselfers.

Why use one particular electronic ac drive over another? This is a hard question, and usually there is more than one answer. Each application dictates what is required. The first issue is deciding whether or not a dc or an ac motor will be used. If we are dealing with a retrofit of an existing motor, then we need to know everything we can about that motor—usually motor nameplate data will suffice. Once this issue has been decided, then the application requirements take precedence. Will there need to be braking or regeneration? What is the horsepower? What is the supply voltage? What speed regulation will the application require? Torque regulation? Is cost a factor? Isn't it always? Is floor or wall space at a premium? Where can we mount the unit? Is a digital or analog drive preferred? What will the home owner find when there is a problem? What kind of duty cycle or loading is predicted? Is the power coming into the home a "clean" voltage supply? Do we need an efficient drive? The questions seem to be endless. Let's look at the issues.

Once it has been decided that an ac motor will be used for the application, we must next decide on the horsepower, voltage, and enclosure, and most important, can the motor selected run properly with any ac drive's output? Is there a good match between motor and drive? Misapplying drives is, was, and will be a potential problem. Many drive misapplications from years ago caused many an engineer and equipment manufacturer to be wary. Those applying drives these days have lived through a bad application or two and want to avoid another. There are those individuals who are going to stick with a certain ac drive technolgy or a certain type of drive. Then there are still others who are willing to push the technology and try something different. Regardless, the application really dictates what to do. Be wary of manufacturers of drives who may have biased reasons to support or degrade a particular design. Listen and learn. Ask the same question of many. The best suggestion to be given is: Take the time and perform an in-depth analysis of the current equipment available. Also, completely understand the application from mechanical, electrical, and connectivity vantage points.

One manufacturer's variable-frequency drive may be well suited for one application but not for another. A variable-frequency drive that is well suited to the constant-torque loads and load changes of a washing machine may be overkill for a swimming pool pump filter application. There are many considerations to ponder when selecting a variable-frequency drive. Section 15, electrical, of architectural specifications usually contains written descriptions of drives to be used on building and construction projects. Custom-built homes that have detailed specifications will have to address this new equipment. Try to secure a copy when purchasing or just to get a better understanding. Talk to as many technical vendors as possible because the industry is constantly changing. They can advise you about whether your application is right for their drive, and vice versa. Here is a list of issues and concerns when ac drives are involved:

Constant torque or variable torque.   How much current is required on a continuous basis for the particular application? What does the speed-versus-torque curve look like? Perhaps the application only needs starting or peak torque for a few seconds; if so, then one frame size of an ac drive may be more suited than another. Remember, electric current will not only affect your monthly electric bill but can be a safety hazard if not addressed properly.

Complexity of the ac drive's circuitry.   The more components, the more risk there is of component failure. Also, consider how long a particular complex drive type has been in production. Where are similar applications in service with the same design? Reliability is critical, and in an ever-changing industry, how long will the chosen design be built and supported?

Digital or analog ac drive control.   While most manufacturer's will promote digital designs with hundreds of functions and features, analog types are usually simpler and less expensive. Digital drives offer much more capability in diagnostics, protection, and communications.

Speed range.   What is the speed range or turndown ratio, and what is the loading like at various speeds? Turndown ratios of 10:1 are common for variable-torque applications. Low speeds while fully loaded are the most difficult for ac drives to control. Similarly, hard-to-start loads may require high breakaway torque. Typically, a drive's slip compensation function can provide adequate adjustment.

Encoder feedback.   Check to see if a drive design is retrofittable with a feedback device to gain better speed regulation. If a module can be added to the drive control circuitry to accept pulses from an encoder, then this information can be used by the drive to correct better to the speed reference.

Auxiliary blower at motor.   When a motor runs at low speeds fully loaded for long periods of time, it must be fitted with an external blower with a small motor. Also, the starter for this blower motor, often just a single-phase device, must be included within the drive package. This auxiliary blower will run constantly regardless of the main motor's speed and blow air continuously over the skin of the main motor.

Braking and regeneration.   Is braking required, or can the motor coast to a rest via the friction in the drive train? What will happen in an emergency-stop situation? The drive has no control when it is not being supplied power and thus cannot commutate. If the drive is not commutating, it is not able to regenerate. Therefore, it is common to use dynamic braking even if a drive can regenerate

Adequate power supply. Some drives are more sensitive than others. Some are also phase sequence sensitive. Determine prior to drive installation if the power supply will be stable. If the supply has dips, frequent outages, and surges, then take necessary precautions at the drive.

Cable sizing. Motor voltage is directly proportional to speed in an ac drive variable-speed application. Therefore, any voltage drop is very noticeable at low speeds. When the motor is running near full speed, a voltage drop in the wire does not really affect the speed.

Location of drive with respect to the motor. Running wire and cable is many times the best alternative rather than locating a drive in a nasty or outdoor environment. Even though special enclosures can be built for these environments, they can be expensive. Also, it is recommended that the drive be as close as possible to the motor.

Proper earth grounding. Ac drive and motor applications requiring three-phase power need to use four-conductor cable. Likewise, in a single-phase system, two conductors with a ground wire should be selected based on the current rating of the drive. All equipment in the motor-drive circuit must be tied to earth ground at one location. The input power to the drive should be from a WYE-configured source. All current will be contained within these four wires and minimize interference on the input. The output wiring should allow for the fourth wire, the ground conductor, to be used as the fixed ground connection between the ac drive's enclosure and the motor itself.

Inrush currents. Ac drives generally will limit the amount of inrush current to the motor to no more than 100 percent. This lengthens the life of the motor while providing softer, smoother starting. When line starting an ac motor, typical inrush currents can approach 600 percent of motor rating.

Replacing a dc drive with an ac drive. Check the horsepower and torque requirements. An identical horsepower ac motor and drive may not be enough to do the job. For example, a center-winder with a buildup ratio may need a much larger ac motor than the previous dc motor. Frequently, a dc motor can be run into the field-weakened speed range, thus keeping its horsepower size down. Again, an ac drive and motor may have to be up-scaled in size to get the proper torque at all operating speeds.

Multiple-motor operation. Does the application require that several motors be run from one ac drive? If so, then a current source drive will have limitations because it is load-dependent. Only a few, if any, motors can be taken off-line when running the entire set. A voltage source drive is more suitable here. One important issue is to apply individual motor thermal overload protectors. The ac drive could inadvertently supply too much current to a smaller motor because the drive does not necessarily know it is running smaller motors. Its electronic overload protection is basically for a motor sized for the drive's rating.

Ground fault and short circuit protection. Does the drive manufacturer require an isolation transformer? How will the drive handle a ground fault or a motor short circuit? Conversely, what happens to the inverter in an open-circuit situation?

Power factor. Most ac drives today have a unity power factor in the 0.98 range. Power factor must be taken into account because of utility penalties. The utility will impose a penalty on demand charges for a low power factor. A drive with a constant power factor throughout its speed range is attractive, especially when it will replace a drive that has a poor power factor.

Single-phase input to a three-phase drive. First, it is not recommended that a three-phase ac drive run a single-phase motor. As for single-phase input to a drive running a three-phase motor, this can be done. However, the drive will have to be derated for the desired output at the selected motor. Consult the drive manufacturer for sizing.

Harmonic distortion or content. Variable-frequency drives create disturbances back to their supply and out to the motor. These are input and output harmonics and are prevalent in devices that convert ac and dc power. Determine what your particular system can tolerate. Usually, the ac drive will have little trouble operating in the harmonic distortion that it has created.

Ac drive efficiency. Ac variable-frequency drive efficiency is just as important as the cost. Make the analysis of each with both the motor and the actual drive. Consider efficiencies at all speeds and mainly for the predicted speeds to be operated. Remember, the variable-frequency drive is supposed to be an energy-saving device.

Cooling. Variable-frequency drives are usually air-cooled, some are water-cooled, and some are air-conditioned. Air-cooled drives tend to be noisier than the others. This is due mainly to the cooling fans inherent to the package. Make sure that the higher decibel levels can be tolerated when introducing external fans for cooling. Water-cooled drive systems involve a pump, piping, and a heat-exchanger system on top of the drive's converter and inverter package. These water-cooled systems, although very efficient when maintained, have the potential for leaks, must be cleaned, and have higher initial costs.

Ventilation. A ventilated, cool ac drive is a drive that will function well and for a long time. When installing any drive, whether it is ac or dc, special attention should be given to the heat generated by the drive. The drive has current running through it. This current will produce heat, and this heat has to go somewhere. It can dissipate naturally if there is a light-duty cycle. However, if loading is heavy, then provisions for cooling or ventilating are necessary. First, take a look at the ambient environment around the drive. Is the room in which the drive will be located warm or hot naturally? What is the temperature on the hottest summer day? Next, is the drive going into an enclosure? Is this enclosure going to be completely sealed? Determine if the drive is going to be heavily loaded. Will it run 24 hours a day or intermittently? Most of the time, ventilation fans, pulling ambient air into the enclosure and up across the drive, will suffice. Figure 9.24 shows typical drive cabinets with different ventilation schemes. Another question to be answered is how clean is the ambient air? If dirty air is brought into the enclosure, then a new potential problem can emerge. Dust and dirt will collect on the drive and virtually suffocate it. Eventually, no heat will be able to escape from the drive, and it will overheat. Most newer drives will trip or

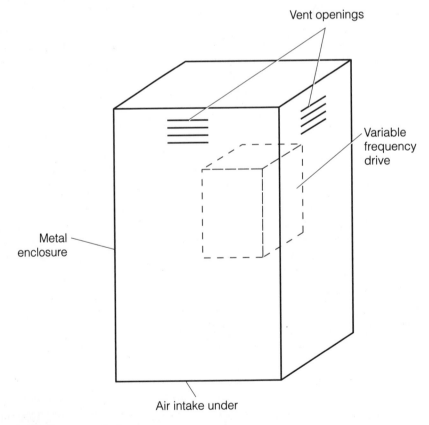

■ **9.24** *Drive ventilation.*

fault on an overtemperature fault. This protects the drive but is a nuisance because the drive will have to cool off before starting it again.

Another means of handling drive-enclosure heat is by air-conditioning the cabinet. This is the more expensive approach but sometimes the only answer. When the ambient air is too warm or too dirty, then air-conditioning makes sense. There are many manufacturers that specialize in small, compact air conditioners that attach directly to the wall of a drive enclosure. These are self-contained, closed-loop units that keep the inside of the drive enclosure completely cooled. Keep in mind that if the ambient area is dirty, the air conditioner will have to be maintained clean in order to operate efficiently. Also, if the air conditioner stops running for whatever reason, the drives will trip quickly because they are now in a completely sealed enclosure with no way for heat to escape. Many air conditioners today are made to work without any chlorofluorocarbons (CFCs). They are "green" products, environmentally safe. However, they will need cooling water supplied to them and will require a drain. They are worth looking into.

High altitude. High-altitude locations can be a problem for a drive installation. Knowing the elevation conditions ahead of time can head off drive-related problems. As the altitude increases, so too does that air's inability to dissipate heat. Thin air, which is what one might get above 3300 ft above sea level (or approximately 1000 m), cannot hold as much heat as an equivalent amount of air at sea level. Therefore, transferring the heat from our heat source, the drive, is more difficult at higher altitudes. Drives often have derating values for equivalent horsepower or continuous current output for a given level of elevation.

High humidity and excessive moisture. Typically, this is a problem for a drive if the excessive moisture in the air condenses on the drive's components. Water will conduct electricity, and if enough water forms on the drive, short circuits can occur. The maximum level of moisture in the air is a relative humidity of 95 percent. This again causes nuisance tripping, and likewise, downtime will occur until the drive is dried. One means of dealing with condensation is to install space heaters within the drive enclosure. These should come on when the drive is off. While the drive is running, enough heat is generated by the drive itself to keep moisture from condensing. Once the drive is shut down, the heat production also will cease. The circuitry should be such that the space heaters should then energize. Some installations with smaller drives can even make use of a standard light bulb as the "space heater."

Another moisture-related problem is that of harmful corrosive gases in the air such as acid mist, chlorine, saltwater mist, hydrogen sulfide, and others. Problems can range from slow deterioration of the printed circuit boards to actual corrosion of the bolts that hold the drive together. Protective coatings to the boards can help, but the better solution is to keep the contaminated air from getting to the drive at all.

Long distances between motor and drive.   This can be a factor depending on the type of drive being used. A voltage-source drive tries to maintain a constant voltage in the motor-drive circuit. Long runs of cable can create voltage drops that can affect the motor's ability to maintain speed under heavier loading. One solution is to increase the gauge or diameter of the cable to minimize drops in voltage. This may have to be done if the actual supply voltage is lower than nominal. For instance, if the supply is supposed to be 460 V, most drives are able to handle a 10 percent range above or below nominal. However, if the supply is lower than the projected 460 V, then a long run of cable can further reduce that value to the motor in fewer available volts per hertz. This may limit speed and torque capabilities. Typical distances, though, are less than 300 ft and usually present no problems.

On occasion, it is necessary to locate the actual electronic drive farther away from the motor than 300 ft. If the drive is a voltage-source drive with very high switching transistorized output, then a different phenomenon can exist. This is sometimes called the *standing-wave condition*. This condition can make for high peak voltages and possibly damage motor windings. This condition is more prevalent with these types of drives and very long runs of cable (usually over 300 ft). One solution is to install output reactors between the motor and the drive. This reactor smoothes out the voltage but adds impedance to the system. This can add to the overall voltage drop and must be considered when evaluating this option as a solution. Again, if the voltage supply is steady and higher than nominal, then all should be fine at most speeds and load conditions. Two other considerations relative to this standing wave condition: (1) the longer the distance between motor and drive, the higher the impedance output reactor to be used, and (2) check the class of insulation in the motor because it may be better able to withstand voltage peaks without substantial degradation.

Input and output contactors.   Often it is such that a motor is located very far from the variable-speed drive, even completely out of sight. If a maintenance person wants to work on the motor, he or she wants to ensure that there is no electricity possibly going out to the motor. One common practice is to install a contactor or disconnect near the motor in the circuit between the output of the drive and the motor. This is fine from a safety point of view but can be potentially harmful to the drive. For example, what often happens is the maintenance person will go onto the roof of a building where a motor is located driving a fan. The variable-speed drive is located two floors below in a mechanical room. The fan is running along at full speed and full load, and the maintenance person decides to "kill" power to the motor by opening the contactor at the motor. This possibly could cause a high-energy spike back to the drive and could blow an output device. Rather than assume that all maintenance personnel are trained to never open an output contactor under load, it is more practical to interlock a contact that faults the drive first and then opens the output contactor. This will ensure that there will be no power flowing through the drive and thus eliminating the possibility of a spike. This contact must be an early auxiliary contact.

Likewise, an input contactor located on the supply side of an ac drive can and should be interlocked to open in certain situations. These situations are any which dictate that the drive should not be able to receive any input voltage or current when a fatal fault has occurred or the drive is attempting to dynamically brake. No new energy is wanted at that time, so the input contactor acts as the shutoff valve and as the disconnect in these situations. Make sure all interlocking is proper.

Input reactors.   Do variable-speed drives require isolation transformers? First, what are you trying to accomplish? Are you trying to minimize noise in and out of the drive at the supply point? Do you need ground fault protection? The isolation transformer can provide all of this, but the home owner should understand the type of drive being supplied before requesting input line reactors.

Output reactors.   Sometimes ringing needs to be reduced at the output to the motor with certain types of drives. Two major reasons, however, to use output reactors is to protect the inverter power devices from a short circuit on the drive's output wiring and to overcome long cable runs from the drive to the motor by smoothing the voltage and adding impedance to the system.

AC Drive Functions and Features

Electronic ac drives today have many built-in features. The power of the ac drive is not only in the converting and inverting sections but also in the software that controls the drive. Early versions of ac drives had to include extra printed circuit boards or plug-in modules to accomplish specialized tasks. Now, with digital technology, high-speed microprocessors, and ample memory, all one has to do is call up the parameter on the drive's display and make the necessary change. Setting up an ac drive is 10 times easier today than it was when dip switches and potentiometers had to be "tweaked." Following is a list of many common ac drive features and functions along with a description.

Acceleration.   This function allows the home owner to select the amount of time desired to reach full speed. This is often referred to as *ramping*, and this function actually provides the soft start capability that limits in-rush current to an ac motor. High-inertia loads may need several seconds, even minutes for some large fans, to accelerate to full speed. If this were not adjustable, then the drive would constantly trip off-line under these conditions. Some drives even can provide two acceleration settings, or a two-stage ramp. As the drive begins to accelerate the motor, a contact closure can signal that it is time to "shift gears" and go into a faster ramp up to speed. Any acceleration setting will be operational even when a speed change is desired. Going from low to high speed will use the set acceleration rate. Special acceleration and deceleration curves can be programmed with many drives. The ever-popular S-ramp for real soft starts is common.

Deceleration.   This is similar to acceleration in setup but most often with a different value. Deceleration is a controlled stop provided by the ac drive. It has limits and is load-dependent. If a high-inertia load is to be brought to a fast stop, then the drive must have some method of handling the motor-generated energy. Regeneration or a dynamic braking circuit will provide an outlet for the brake energy. If not, then the drive can provide some braking power by allowing its dc bus voltage to rise slowly. If this voltage rises to the maximum set level (set there to protect devices), then the drive will trip on an overvoltage fault. The solution here is to try a longer deceleration rate.

Automatic restart.   Some ac drives come equipped with the ability to automatically attempt to restart themselves when conditions permit. For instance, a drive can be programmed to automatically attempt three times, at 2-minute intervals, to restart. As long as power is available to the control circuitry, the drive's logic unit will allow the restart attempts. If after the selected number of attempts the drive cannot restart, then it remains in a faulted condition until someone resets it manually. Frequently, a manual reset is preferred for safety reasons. We may want to know that the drive has shut down so that we can diagnose the situation. A typical automatic restart function is appropriate whenever supply power is known to dip below acceptable levels of voltage (undervoltage).

Loss of automatic signal.   Many drives can be programmed to run manually from a potentiometer or keypad setting right at the drive. Other times it is desirable to receive a 0- to 10-V or 4- to 20-mA signal to scale as the speed range. In this situation, 4 mA will equal minimum speed, and 20 mA will equal maximum speed. These minimum and maximum values are also programmable for the particular application. When running in automatic mode, it is necessary to have a safety built in that handles those conditions where the automatic reference is zero (absent) or loaded with electrical noise and outside the min/max settings. Most often the drive will go to a preset safe, slow speed under these circumstances rather than allow the motor to run away. Some drives can be programmed to also shut down and fault, whereas other applications may need to have the motor keep running while a fault is announced. Likewise, many ac drives that may be accepting an encoder or tachometer feedback signal may need to fall back to a safe, slow speed if that signal is lost. Look into this ahead of time.

Skip critical frequencies.   In many drive applications, especially fan and pump applications, there is always the possibility that resonant frequencies can exist. And many times these resonant frequencies can cause severe vibration in the mechanical drive train of the drive and motor system. If the drive were set at this frequency and run continuously there, then possible premature mechanical failure could occur. The ac drive can be programmed to avoid certain frequencies. It will select a frequency above or below a set bandwidth in order to skip the known resonant frequency. This is illustrated in Fig. 9.25.

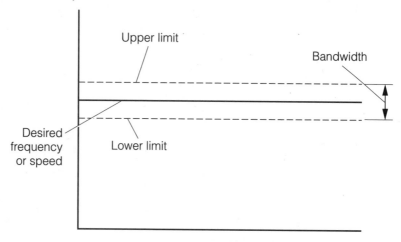

■ **9.25** *Bandwidth for critical frequency avoidance.*

Fault logging.   Many digital drives are kind enough to display the fault that has taken them out of service. Many are not. Some restart automatically and keep running; the home owners are never aware of any problem. Many ac drives have enough memory to store faults as they occur and thus have a record of when a fault happened, what it was, and how the drive was reset (automatically or manually). Figure 9.26 shows a typical fault log. Usually these faults are stored in a first-in, first-out manner.

Power loss ride through.   An ac drive has the ability to ride through short power interruptions. The drive can continue to run on stored dc bus energy until the bus can no longer supply enough voltage to keep the drive running, and it then faults. Most drives can only ride through 1 to 2 seconds.

Slip compensation.   The amount of slip in an ac motor system is proportional to load. With increased slip comes an increase in the necessary

Fault history log for a 1 Hp, 115 Volt, Single phase, 60 Hz, heat pump VFD

Fault#	Fault type	Reset method	Input voltage	Output current	Time	Day
1	Fuse failure	Manual	128	4.3	1351	6/22
2	Fuse failure	Manual	119	3.7	1355	6/22
3	Undervoltage	Auto	92	1.3	0930	7/23
4	Overvoltage	Auto	133	1.9	0112	7/23
5	Overtemp	Manual	115	1.8	0915	8/19
6	Overload	Manual	129	2.9	1115	9/03
7	Overload	Manual	128	2.9	1125	9/03

■ **9.26** *A variable-frequency drive fault log.*

torque to continue driving the load. However, often the speed of the rotor actually slows in order to continue driving the load. In applications where this speed shedding is undesirable, slip compensation provides a solution. As the load increases, the ac drive can automatically increase the output frequency to continue providing motor slip without a decrease in speed. The actual amount of slip compensation will be proportional to the increase in load. Usually one setting of slip compensation will cover the entire operating range with speed regulations of 0.5 percent.

Pickup a spinning load.   Many ac drives are expected to catch a motor spinning with load at some speed and direction and take it to the actual commanded speed and direction. This means that the drive has to be able to determine the present motor speed instantaneously, determine the direction, and then begin its output at that speed. The drive then has to reaccelerate or decelerate the motor to the desired reference speed. Applications with large, high-inertia fans are good applications for this function. A large fan blade can take several minutes to come to a rest after running full speed. The ac drive can start right into this coasting motor.

Motor stall prevention.   There are some instances where a motor can get into a temporary overload condition. The drive's normal operation wants to protect the motor and shut down. This can be a nuisance fault and may not be acceptable. Many fan systems that begin moving cold air experience these overload conditions. However, once the air warms, the motor becomes less loaded and will continue to operate as required. This function allows the drive to lower the output frequency until the output current levels begin to decrease. In this way the drive will ride through the overload condition without tripping.

Maintenance of AC Drives

Many new installations of ac drives within the home and in various appliances initially will embed the drive inside the equipment. The drives are embedded in such a way that the average home owner will not be able to get at the drive to know if there is a problem. A service person will have to be called at rates of $30 to $80 per hour. However, for the tinkerer and do-it-yourselfer, here is some good information:

> **NOTE:** *All input power must be removed from any ac drive before doing any maintenance. Also, live voltages can exist in drive capacitors. Check to see that all energy has dissipated.* Also, if there is a lockout mechanism in place on the input power line, it should be used before doing any maintenance.

A routine inspection should be done on an ac drive every 3 to 4 months, and this frequency should be sooner if the environment around the drive is very dirty. The inspection and subsequent maintenance should include the following:

1. Visually check the drive for any dust buildup, corroded components, and loose connections.

2. Using a vacuum cleaner with a plastic nozzle to minimize damage to components while cleaning, brush dusty and dirty devices while vacuuming. Particular attention should be paid to the rectifier heat-sink areas. Excessive accumulation of dirt and dust eventually will lead to an adverse overheating condition (probably at a time of peak production). One must clean the drive to avoid these types of problems.

3. Clean and retighten any loose electrical connections.

    a. All fuse ends and bus bar connections should be cleaned according to the manufacturer's instructions.

    b. In the case of screw pressure connectors, the power connections should be checked and retightened, if necessary, often during the first few weeks after installation.

    c. If any components are unfastened during the maintenance procedure, then the mating surfaces should be treated with the proper joint compound (per the manufacturer's recommendation) before replacing. After replacing any power modules subject to certain torque values to their heat sinks, it is suggested that another check be made on the tightness of the screws.

4. Touch up any exposed or corroded components with paint.

5. If the ac drive is mounted within an enclosure, check the fans and filters. If necessary, clean or replace the filters and replace the fan if the shaft does not spin freely.

Troubleshooting and Repair of AC Drives

Many ac drives are manufactured in a similar fashion to one another. Surface mount device (SMD) technology has helped make drive circuit board integrity very good, and many drive manufacturers are using this technique. Basic power conversion, available devices used, and standard wiring and protection techniques also can be assumed to be similar from manufacturer to manufacturer. However, control schemes, gate firing circuits, packaging, and so on will definitely vary from drive to drive. Thus it is mandatory that the manufacturer's installation, maintenance, and repair manuals be relied on to perform any work on ac drives. Especially with digital drives and the numerous parameters that can be set and the extensive diagnostics provided, the manual will be the only place that will define many of the displays. This section will discuss standard and common techniques along with practical applications and problem solving. *Again, a warning about working on electrical equipment must be provided continually. Electric shock can cause serious injury or even death. Remove all incoming power before attempting to do any work to any electrical device!* This is another reason to have the manufacturer's manuals on hand, since they usually can take the repair person step-by-step safely through the situation. Whenever possible, work with another individual. Also, when using an oscilloscope, be aware that it, too, must be properly grounded.

Many times in the "heat of the moment," when the drive has quit running, it is usually much faster to just replace the drive for another of equal size and rating. Hopefully, there will be modules, boards, and plenty of fuses on hand, at least with the business from which the drive and equipment were first purchased. If not, then shame on the personnel who decided that this issue was not important. Spare parts usually can be overnighted from somewhere in the country, but nothing is better than to disconnect the bad device (work that has to be done even in a repair situation) and get back up and running.

Every manufacturer of ac drives will have a recommended spares list of those parts which should be kept on hand in the event of drive failure. The best time to purchase these parts is at the time of drive purchase. This should be the moment when the buyer has his or her best buying power. Because later the prices of the spare parts become two to three times what they should have been. Besides not having them when they are needed most, we end up paying much more for them, and the overnight air freight is another killer!

For a typical ac drive, the following, at a minimum, should be on hand: various fuses at the ratings seen throughout the drive. Input fuses and branch fuses are a must because they protect the bulk of the drive's components from incoming power surges. These fuses do not have to be purchased through the drive supplier as long as the current ratings and sizes for the recommended fuses are matched. The power modules, transistors, or SCRs (Silicon Controlled Rectifiers) typically are not items that the average home owner is going to replace or for that matter the service technician either. A complete drive will be swapped out rather than spending the time to take the bad unit apart and find the damaged components. The devices may have certain turn-on/turn-off values that may cause problems later if not matched accordingly. Often, when an SCR or transistor fails, it is common practice to replace the other devices in the leg of the circuit because they may have become stressed and may be prone to quick failure.

Other components that should be on hand as spares are the printed circuit boards for the drive. Unfortunately, these boards are not going to be bought anywhere else but from the drive manufacturer. The control boards and gate-firing boards are the most common to stock, but consult with the drive manufacturer for any others. Also, consider electrostatic discharge (ESD) when handling any printed circuit boards. A grounding wrist strap is ideal but rarely practical when servicing a drive that is out of service. The next best procedure is to have an ESD-protected bag available, minimize handling of the board(s), and do not touch small components on the board. Use one hand whenever possible to keep a path for voltage arcing from forming, and get the board into the bag as quickly as possible.

There may be some suppression components in place to minimize further component damage in the event of surges, and these can be stocked as spares but can be found in many electrical hardware supply stores. They are the capacitors and resistors in the snubber network. Power supplies are other items that sometimes fail and are usually not manufactured by the drive supplier but rather purchased elsewhere. Determine whether or not it is worth keeping these items in stock.

Finally, to round out the necessary spare parts in the drive's inventory, any fans, feedback devices, special boards (for feedback, etc.), and temperature sensors probably should be kept on hand if the drive is to be kept running 24 hours per day, 7 days per week. And as another precautionary measure, continue to learn as much as possible about how to troubleshoot, replace, and repair the drive components. This training must be specific to the drive in place and usually can be performed by the drive supplier.

Whenever the drive manufacturer can provide the field service, training, and additional support, it is always practical to get these services. However, these services are usually very expensive, and not everyone in the house can get to be an expert on every drive. This is one reason why manufacturers of equipment using ac variable-frequency drives try to standardize on one or two particular drive manufacturers. However, this methodology is not always absolute because drive manufacturers change their designs over time and home owners must relearn the product anyhow. The following sections will attempt to cover the common problems, faults, and diagnostics that are inherent to most ac drives and their applications. Many drives have onboard diagnostics that can help greatly in tracking problems. Sometimes this is in the form of a display either in letters or by a numeric code. The drive manual, which should be stored in a pocket on the drive door and not in the garage or kitchen closet, should be consulted.

When a drive is down, the first thing to do is to kill incoming power and lock it out. Be wary of drives that contain capacitors, which can carry charges for a period of time. Many drives carry a LED that indicates, when it's on, that the capacitor charge has not been fully bled off. *Be careful!* On opening the drive cover or door, the initial checks should be to look for any apparent signs of internal component damage. That smell of fresh electrically produced smoke is a good sign. Next, look for any burnt wiring or components. Some components may even swell, so look for these, especially the capacitors. Any loose or disconnected wires should be noted, and any marks on the drive enclosure walls from apparent electrical arcs also should be noted. Once the visual inspection is done, then begin further troubleshooting and tracing from the suspected point.

Check all input fuses to see if any have blown. This may be apparent visually, or a physical check with an ohmmeter may be required. Either way it must be determined as to whether the problem lies before, at, or after the fuse section. Frequently, the fuses are blown, and there is still a second or third problem with the drive. This is where the fun of troubleshooting begins. Again, some drives may have the capability of saving multiple trips and faults. Use this tool.

Since most ac drives contain diodes and transistors, it is common to suspect these devices whenever there is an apparent problem. Many drives have the capability to display, when faulted, which transistor has failed to turn on or has been forced off abnormally by excess current. One probable cause might be that a high-inertia load has been accelerated too quickly. In this case, merely setting a longer acceleration time will solve the problem. Another cause could be the voltage boost. While this is a nice tool for hard-to-start loads, it is also a sensitive tool. Reducing its value will overcome motor losses and keep the transistor from failing. The last probable cause is certainly not the easiest to correct. This would be a failed transistor.

In order to determine if a transistor module is faulty, we have to measure the resistance across its terminals with an ohmmeter. Measuring normally with the one times range we can get readings that, when compared, with the manufacturer's acceptable and nonacceptable levels, will indicate whether the transistor needs replacing. Transistors commonly come in modular form and are fairly simple to remove and replace. The same procedures should be followed for any of the diodes.

Common Drive Faults, Possible Causes, and Solutions

Electronic ac drives today tell us much more than the drives of yesterday. When the drive trips, we can usually get a self-diagnosis from the unit. Some go so far as to provide actual text on a terminal screen that tells the troubleshooter where to troubleshoot. Many drives trip or fault for the same reason, but the solution may mean looking at a certain terminal or adjusting a certain parameter for that particular drive. Therefore, the specific drive manual must be referred to for this kind of information. The common drive trips, sometimes even called *alarms* or *errors*, are as follows:

1. The motor will not run.

   Possible cause: No line supply power; output voltage from ac drive too low; stop command present; no run or enable command; some other permissive not allowing drive to run; drive could be faulty.

   Solutions: Check circuit breakers and contactors on input power side.

Check to see if stop command is absent when trying to run motor. Check for run, enable, or other permissives being present. Try another ac drive if the one in place does not respond.

2. Low line input, undervoltage, or power dip.

   Possible cause: Input fuse(s) blown; low input voltage; power outage.

   Solutions: Check and replace fuse if blown; check to see that input voltage is within drive controller's allowable high and low range; check for possible problems with utility; sometimes lightening can cause this fault.

3. Overtemperature.

   Possible cause: Heat-sink temperature sensor has caused trip; cooling fan has failed; drive ambient temperature has risen to extreme levels.

   Solutions: Check thermistor with ohmmeter; if bad, replace. Are heat sinks dusty and dirty? They should be clean. If drive is in an enclosure, are there filters? Are they clean? If not, clean or replace them. Is there a cooling fan? Is it operational? If not, then repair it.

4. Overcurrent or sustained overload.

Possible cause: Incorrect overload setting; motor cannot turn or is overloaded.

Solutions: Check overload parameters, and change to fit motor loading; load may be too large for motor (a larger motor may be required, or the gearing may have to be changed); if there's a jam in the drive train, clear it.

5. Motor stalls or transistor trip occurs.

Possible causes: Acceleration time may be too short; high-inertia load; special motor.

Solutions: Lengthen acceleration time; readjust volts-per-hertz pattern for application.

6. Short circuit or earth leakage fault.

Possible causes: Motor may have a short circuit; power supply may have failed; excess moisture in inverter or motor.

Solutions: Check motor and cables for short; have motor repaired; replace power supply.

7. Overvoltage.

Possible causes: Dc bus voltage has reached too high a level.

Solutions: Too short a deceleration time for load inertia; supply voltage too high; motor may be overhauled by load.

8. Main fuse(s) have blown.

Possible causes: Bad transistor(s) or diode(s); short or bad connection in power circuitry.

Solutions: Check devices and replace if found defective; trace and check connections in power circuitry.

9. Speed at motor is not correct; speed is fluctuating.

Possible causes: Speed reference is not correct; speed reference may be carrying interference; speed scaling is set up incorrectly.

Solutions: Ensure that speed reference signal is correct, isolated, and free of any noise; filter noise on-line and/or eliminate noise; recalibrate speed scaling circuit.

10. Peripheral relays or control communication circuits tripping.

Possible causes: Drive carrier frequency too high; other electrical noise present, excessive panel heating.

Solutions: Lower carrier frequency; eliminate electrical noise.

There are certainly other types of faults that can be seen for any given ac drive and motor installation. Digital ac drives can always contribute the ever-deadly CPU failure or show no response at all when the main microprocessor board is not functional. Sometimes this is also called a *watchdog timer fault* with the main control board. Also with digital drives, if the drive can tune itself automatically with the motor and an operational error is detected, then the autotune sequence is aborted and an alarm is made. Additionally, any drive controllers that are set up to accept other external feedback, and it is not present, may trip on powering up. As for motors, they can overspeed or run away, but usually the drive controller can be set up to trip at a maximum speed for protection.

Applications for AC Variable-Frequency Drives within the Home

Whether or not the home owner accepts this technology does not really matter. Manufacturers of equipment used in the home will make the change for us. Washing machine makers and heating and cooling equipment manufacturers want an edge. Therefore, if their product with this new energy-saving technology helps to gain market share over their competitors, then they will find ways to incorporate it. As costs of ac electronic drives come down, their application becomes more justifiable for use in residential appliances. Additionally, new concepts for energy savings can be tried for their practicality.

The furnace fan in the home typically consumes almost twice as much energy as does the total lighting within the house. The furnace fan running 10 hours per day with many, many starts and stops can consume over 100 kWh of electricity in a month's time. A variable-frequency drive placed on this motor would help in many respects. First, any reduction in motor (fan) speed would save energy. This is proved by the fan and affinity laws of physics. Second, as the ac motor starts, and this is typical of any ac motor under load, the electric current draw is three to four times the value usually seen while running. This inrush current takes its toll on motor windings over time. Envision 30 to 40 starts per day times 7 days per week—that's nearly 300 starts every week. Motors do not last forever, and they usually fail at the most inopportune time (on the coldest day of the year).

As can be seen in the line diagram in Fig. 9.27, a variable-frequency drive could be implemented without much trouble at all. Turning down the speed just a little can save energy dollars right away. The ac drive also will limit the inrush current to the motor, thus softly starting it and lengthening its overall life. Additionally, there have been some studies that suggest that the furnace fan should run all the time. If this were common practice and the motor ran full speed all the time, the furnace fan costs could soar to

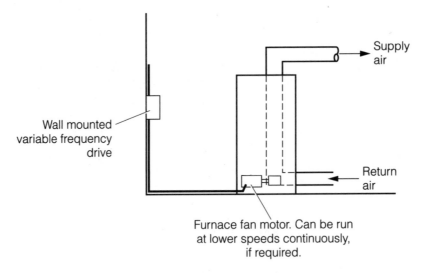

Wall mounted variable frequency drive

Supply air

Return air

Furnace fan motor. Can be run at lower speeds continuously, if required.

■ **9.27** *Drive in a home fan system.*

around 25 to 30 percent of the total monthly electric bill. This would not be acceptable to most people, and this perhaps explains why this practice has not been implemented. Keeping air flowing throughout the house can keep cold and hot spots from forming and, with a good filter system in place at the furnace, keep the air fresher and cleaner. However, energy costs keep this from being practiced—until an ac drive is considered!

An ac drive can keep a slow, steady speed at the fan motor when heat and cooling do not have to be delivered. This slower speed will use only the energy needed to move a small amount of air—thus making the actual operating costs practical. Whenever there is demand for full heat or cooling from the thermostat, the variable-frequency drive can be programmed to change to a higher speed until the thermostat is satisfied and then drop the fan motor back to a slow speed. The drive allows for virtually any speed setting that the home owner desires or needs for his or her comfort level. Ironically, the energy costs associated with this mode of operation will be less than they were for starting and stopping a motor 300 times per week. The initial cost of the ac variable-frequency drive and its installation shall have to be factored in when justifying this application.

Another application of ac variable-frequency drives already being implemented is that of optimal heat-pump system control. The performance of the heat pump can be enhanced to make the system more efficient. A typical heat-pump system is shown in Fig. 9.28. Notoriously, the heat pump is not for every climate. It works in one cycle to heat a residence and reverses that cycle to cool the residence. The main drawback is that whenever the temperature outside is too low, not enough heat exchange can be made fast enough to satisfy the thermostat and eventually the occupants. This is

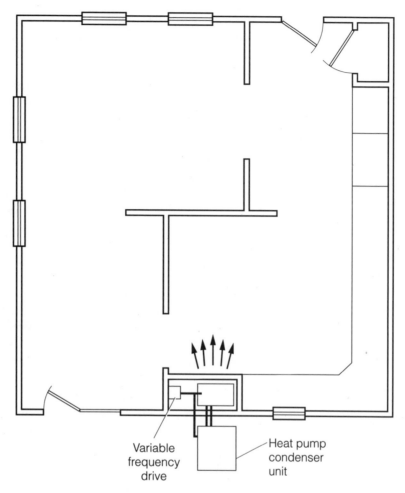

Variable
frequency
drive

Heat pump
condenser
unit

■ **9.28** *Heat-pump application involving an ac drive.*

when the high-electricity-consuming coils come on as auxiliary heat. This is good, quick heat, but it is extremely expensive. The ac drive cannot correct this condition, but it can smooth out motor operation during the many other instances where temperatures are milder. In doing this, energy savings can be attained.

A typical heat-pump system must have several circuit breakers to function properly and safely. Sometimes two 30-A and one 40-A breakers are required for a heat-pump system. This is a possible total of 100 A of service that can be demanded, and this is why a residence with a heat-pump system must have a minimum of 200-A service to the house. As can be seen in Fig. 9.28, there is a fan and motor in the outside unit for air to air heat exchange. There is also a fan motor attached to the inside unit to move air within the house and across the coils. There has to be a small compressor motor to compress the refrigerant to move heat from place to place, and lastly, there are those dreaded electric coils for auxiliary heat. This unit has the potential to consume a large quantity of electricity, and this is why the electrical heating and cooling loads of any residence are the greatest portion of any monthly electric bill. Until the heat-pump technology is antiquated and eventually replaced, heat-pump manufacturers are continually striving to make the heat-pump system as efficient as possible. This is where the variable-frequency drive can come into the picture.

Whenever the home is fortunate enough to have a swimming pool or spa on the property, there are always going to be pump motors needing control. In the case of the swimming pool, this is usually the pool's filtration system. Sometimes it is desirable only to run the filter at a slower rate rather than full filter. An ac variable-frequency drive may be used in this instance. With a jacuzzi or spa, the water jets must have the ability to spray faster or slower depending on how many people are in the spa and what the desired jet action is. Variable-frequency drives provide exact control of these systems. Other applications of ac variable-frequency drives within the home are on some types of treadmills, whereby the speed of the tread can be adjusted by the user, and on in-house rehabilitation and water aerobic systems, which allow the adjustment of water current by the user. Thus it can be seen that ac variable-frequency drives are quickly integrating themselves into the residential sector. Refrigeration, freezer, and compressor manufacturers are looking at this technology further to find justification to implement the drive on their equipment. As the costs of these variable-frequency controllers continue to come down, justification will be made very easy. As time goes on, more and more differing applications will be fitted with ac variable-frequency drives within the home.

Applying Electronic AC Drives in the Home

Electronic ac drives have come down so much in cost over the past 10 years that the residential marketplace is now poised to accept them in quantity. The ac drive can provide energy savings, and if a rebate program is offered by the electric company in the region, then home owners can take advantage of it. If energy savings are not a good enough reason, then is the motor protected. An ac drive will protect the motor from overloads and will keep it from burning itself up. Another attractive feature that ac drives provide is soft starting. This type of starting can add years of life to the ac motor and also can save wear and tear on the mechanical system to which the motor is attached. Another benefit! Does the application require that the speed be lowered? We have shown that the best way to accomplish this with an ac induction motor is by implementing an electronic ac drive. This is the quickest and smartest way to vary motor speed. Thus there are many very good reasons to incorporate an ac drive with an ac motor these days. With surface-mount technology and several drive competitors vying for all this drive business, the home owner is presently in the "driver's" seat!

Because so many ac motors are in place, and because of the fact that ac drives can provide speed control for a standard NEMA design B squirrel cage ac motor, the home owner should look at the ac drive closely. This is attractive obviously for new installations but also for retrofit installations. Someday in the not to distant future we will be able to purchase these ac variable-frequency drives at the local hardware store. We can already purchase the ac motors when we have to. It is also less expensive to purchase an ac squirrel cage motor. It will cost less than a similar-horsepower dc motor because it has no brushes or commutator. It is also smaller in size for a given horsepower, thus making its metal content lower and again effectively making it less expensive. A motor's enemy is heat buildup, and there must be adequate means of dissipating the heat. Sometimes auxiliary fans must be used to provide cooling. These fans are usually cycled on and off by the ac drive. Motor heating must always be considered, as much as it affects motor construction, which affects the cost.

Multiple ac motors can be run simultaneously from one ac drive sized to handle the total full load currents of all the motors, as seen in Fig. 9.29. You could have two or more motors all running at the same speed, slow or fast. Therefore, if you had multiple fans installed around the house, then each fan motor could be wired to the single drive. This arrangement would be cost-effective and would yield good energy savings.

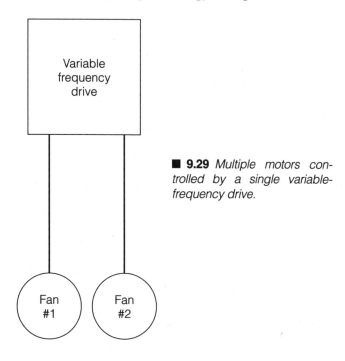

■ **9.29** *Multiple motors controlled by a single variable-frequency drive.*

Does the introduction of an ac drive, another electronic component to the system, increase the likelihood that your furnace or water pump may not work if the ac drive quits working? The answer is no because another major feature of the ac drive is its ability to be bypassed in the event that there is a problem with the drive. Since an ac motor is being used and may have run full voltage and at line frequency (60 Hz) prior to the ac drive, a bypass system can be installed with the ac drive or even later. In an emergency, the ac motor can be run at full speed, as is shown in the bypass scheme in Fig. 9.30. There have to be contactors in place to isolate the drive from the supply line and the motor whenever the motor is running in bypass, and the converse is true whenever the motor is running off the ac drive. Also, motor overload protectors are recommended whenever the ac drive is not in control, electronically protecting the motor.

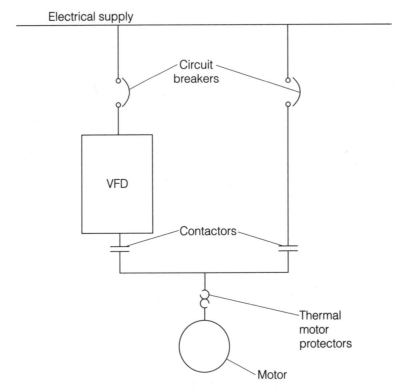

■ **9.30** *A constant-speed bypass system to ensure emergency operation.*

Electronic ac drives can pay for themselves in energy savings, protect our ac motors and other drive train components from excessive wear, and give us automatic control that we never had before. Installing the ac drive is also not as troublesome as it once was. Home owners eventually will gain the know-how to maintain and troubleshoot these products working their way into the residence. There also has been an excellent historical base of ac drives in service for many years without incident. The reliability and integrity of the product are very good. And since more people understand them better, misapplying ac drives has diminished. In the future, ac drives will be even more affordable, and residential use will increase dramatically. Look for them in your local hardware store someday!

Residential Impact of AC Variable-Frequency Drives

Can we foresee into the not to distant future the need for a homemaker to reset an ac drive on her washing machine? You bet! The introduction of ac variable-frequency drives into the residential market has already begun. Washing machines, heat pumps, and in-house exercise pools have variable-speed drives built into the equipment today. Home owners and domestic homemakers will all have to educate themselves on this technology. Technology has been forcing itself on us for years. We have to learn different computer operating systems over and over, train ourselves on different application programs for work and home, and learn much more about electricity and electronics. The trend today is for manufacturers of electronic equipment to charge for support and service after an initial grace period. This can be rather expensive and not attractive to the home owner or owner of the high-tech equipment. We are being forced to be self-trained technicians in order to effectively run our homes within a budget.

Another facet of concern with ac variable-frequency drives is that involved with electric cars. While this book's energy-savings theme is geared toward the home, electric cars are an emerging technology that uses the inverter portion of the variable-frequency drive to run, control, and operate an automobile. The switch from gasoline- and diesel-burning engines to electric motors is an energy-savings and fossil-fuel-saving venture unto itself. How this affects the home is interesting.

First of all, the electric car's construction basically is an ac electric motor, dc power stored in several rechargeable batteries, and an inverter that will invert the dc into variable-frequency output to the ac motor, yielding the various speeds. A machine such as this will have to be parked at night in the home's garage and have its batteries recharged. This means that the garage of the future will have to have perhaps one or two dedicated circuits to allow the car(s) to be connected. Special cabling will have to be in place, and more than likely, timers and other controls for this process will be necessary. This extra electricity draw on the home's system also may be monitored separately by the meter and utility in order to track usage. Obviously, large oil and gasoline producers may see sales drop dramatically from the use of electric cars and may strive to find a way to recoup electrical energy dollars.

Ac variable-frequency drives also lend themselves well to any home automation scheme. If the residence is going to have many "smart" appliances and devices hosted to a master controller or computer, then the digital ac drive can fit right in. The digital ac drives have several low-voltage inputs and outputs that can be used to start and stop a motor from a remote location or computer. The desired speed also can be downloaded to the drive from the host as a voltage or current signal proportional as to how fast the motor should run. Logging of fault and running data is also achieved by having a communications link from the digital ac drive to the host controller. From here the equipment can be linked to locations outside the house via telecommunications lines. The digital ac variable-frequency drive has all the features already built in.

Residential Automation

10

We all do it! We physically turn lights off when they are not in use. We turn the thermostat down on the furnace whenever it is warm. We are constantly striving for energy savings by being the "energy cop." Sometimes, however, we are not around when the light is on or the furnace needs to be adjusted. We also make mistakes, are not that precise, and many times do not think of these matters. This necessitates the use of technology and controls that are available today. Residential automation is evolving quickly to a point where it is more affordable, more practical, and very interesting.

Many options are available to the home builder and home owner when it comes to residential automation. With more and more homes equipped with personal computers, familiarity with computer technology is not a problem. Moreover, the tie-in is made somewhat simpler. Additionally, the energy savings that can be achieved through an intelligent monitoring system can vary tremendously from residence to residence. It could be very surprising to many as to how much electrical and heat energy is truly being wasted. In other instances, the cost of an intelligent home automation system could never be justified on energy savings alone. Tying in security, audiovisual, and other functions can make the whole installation worthwhile. The bottom line is that almost everyone strives to save and conserve energy, but sometimes we do not realize where and when that energy is being wasted. An automatic system will do this for us. Besides, every little bit of savings *will help*. And as costs for these systems continue to come down, the technology becomes more of a standard, and the demand is there, more and more homes will have these kinds of systems.

Overview of Home Automation Technology

Over the years, automation has been sneaking up on us in many forms. On the energy side, the room thermostat, which starts and stops the furnace automatically, was implemented many years ago. This thermostat took our place—having to feed the fire or turn on the furnace. Additionally, as more and more appliances and electrical items have been added to the standard home, the total electrical consumption has gone up, thus increasing the demand for electrical service. In the last 15 years, development of the microprocessor has really opened the door to more automation. Thousands of decisions that we had to make in the past are now being made by the microprocessor, and with all the other elements of an electronics system in place, we have the makings for an automated house. Extensive wiring of many types, programmable outlets, master controllers, microprocessors in most every appliance, and plenty of application software programs are bringing the future to us today. The premise for the basic operation of residential automation is shown in Fig. 10.1.

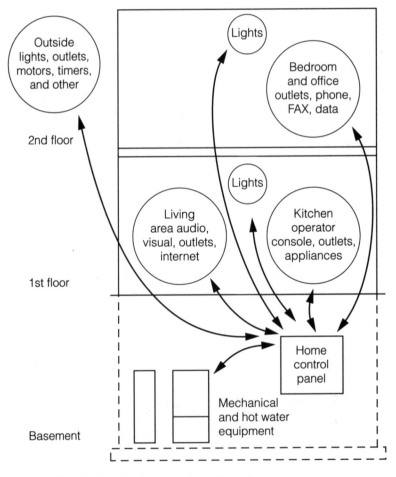

■ **10.1** *The basics of a home automation scheme.*

Electronic Homes

The future is here. Although the Jetsons did not spend too much time discussing their energy needs and energy systems, they certainly were on the right track with the technology. Today, much of the technology exists to transform a residence into an "electronically controlled house." With time, costs will come down, more powerful controller systems will be available, and application software will be written.

Whenever one talks about the future and home automation, the subject makes its way to "future-proofing" and building in of hardware for upgrades. The same subject presents itself with older homes and homes that have not been wired or for which plans have not been made for easy future upgrades. Examples of this include the placing of special conduit in walls and between floors so that the dedicated cables and wires necessary for automatic control can be routed. Likewise, adding underground conduit for outside sensors and associated equipment should be considered. This is the biggest problem confronting the home automation industry. When building a new home and not putting a complete automation system in at that time, the costs associated with any extra wiring, conduit, or hardware are extremely difficult to justify. Many home builders have recognized this trend of homes needing more wiring in the future. One approach is to place one or two runs of plastic pipe into the superstructure of the residence during the building process. These runs, as shown in Fig. 10.2, have a length of

■ **10.2** *Three-inch plastic pipe routed from basement to attic for future wiring.*
(Courtesy of Wayne J. Henchar Custom Homes.)

plastic pipe extending from the basement to the attic. In doing this, whenever the future need for a wire or two or three is there, a path to run the wire is already in place. This will minimize the tearing apart of walls, floors, and ceilings to run new wire.

Even though it is very probable that some new additional length of wire will have to be routed in a residence over its life (for speakers, new lights, etc.), putting it in when the walls are exposed and the installation would be simple is not usually considered. The biggest reason is the extra costs involved for both labor and materials. History is showing us today that the wiring in our homes now is not going to be adequate in the future. Ironically, it costs a whole lot more to install the same wires in later. Therefore, "future-proofing" the residence is a practical endeavor to consider at this time.

Then there is the argument of wireless versus hard wire. As we are being bombarded (literally) with all these high-tech, digital and analog wireless devices, there are still shortcomings to deferring the future of automation to wireless systems. The reasons are as follows: First of all, this technology is rapidly changing. Airspace and allowable frequencies are constantly being sought after and gobbled up by the companies in that business. Graphics and video transmissions involve the transfer of huge amounts of data. Even on dedicated wire systems and modems, the transmission can take a very long time. Wireless systems do not have the vast amounts of airspace needed at present. Sacrifices to transmission speeds and signal integrity could be incurred. Electrical noise is also still an issue with wireless systems (have you listened to a hard-to-hear cell phone call lately). Coaxial and dedicated hard-wired systems are still the best but obviously have to be factored into an initial building scheme.

The costs associated with a full wireless installation versus a retrofit installation also have to be weighed. Likewise, the costs of putting some extra wiring or conduit in when the building is under construction have to be considered. Some forethought as to whether or not future plans will call for selected control of outlets or of a room's electronic functions has to be discussed somewhere in the planning cycle.

Energy Systems: How and What to Control

Many years ago homes were heated by fireplaces, coal or oil furnaces, and even by natural gas. Lighting was accomplished with candles, gas lights, or kerosene lanterns. From that point, electricity and its possibilities emerged on the scene. Electric lights and the installation of a few electrical outlets were now the accepted norm for building construction. At that time, 100-A service was considered almost overkill for home electrical needs. Fuse boxes were the standard, and the electrical future was not really a consideration when building a new home. How times have changed! And they will keep changing.

We have doubled our electrical needs over the past couple of decades, and this trend is not going to change. We have shifted from the electrical to the electronic basis for our everyday life. This has added demand to the electrical needs of every household. Even though the voltage and current levels are lower with electronic equipment, there is so much more of it being integrated into the home every day that the loads are increasing. The electrical utilities are also at a quandary as to what to do. It is not easy to plan, budget, build, and place a new power plant on line. Thus new emphasis is being placed on better use of the electricity that is being produced today. This is also true of *all* energy resources—coal, oil, natural gas, wood, and others—that are not easily replaced. Discussion of utility rebates, peak versus nonpeak usage, and other conservation systems has been provided in earlier chapters.

With the increase of electrical content in our homes, appliances, and equipment, the service to the home has increased. The need for 200-A service is now a minimum for an average-sized family. Each room in the house has to have a duplex electrical outlet every 6 ft of linear wall space (or less). It seems as though the outlets *all* eventually get used, and the need for extension cords becomes an unwelcome reality. When will enough outlets be enough? This is a hard question to answer, and if one looks at the evolution of electrical usage in the home, the answer is still far-reaching. The good news is that this evolution is leading into an approach that will allow for better control of many energy-related functions in the home.

Home Energy Management

Now that we have determined to allow the electronic systems in the house to take charge, we must next decide what energy items need to be controlled. They can be prioritized, but the list will differ for each household. However, the methodology is more critical. The refrigerator/freezers, electric heat (if any), electric clothes dryers, and hot water tanks need to be controlled. Deciding on what to control and how it is to be controlled is the first problem. The "how" part of the equation is broken into several distinct groups with definitions as follows:

1. *Load management.* Load management can be expressed as the electric current demand of appliances and equipment. Within the house, each electrical device has a current rating, usually given as a maximum value in amperes. Therefore, load is synonymous with electric current. Certain appliances and electrical equipment have higher current draw than others. For instance, a heat-pump system with a fan, compressor, and electric coils can draw 40 to 50 A when the weather is severe. This will spin the electric meter furiously, thus driving the kilowatthour usage up quickly. An electrical item such as this obviously should be red flagged as an item to be controlled under a load-management scheme.

Loads within the house can be broken down into different types. There are the traditional on and off states of many devices. Lights, hair dryers, toasters, and electric heaters are examples of on/off devices, each with a defined amount and effect to the current draw from the electrical supply. Another type of load is the multiple-stage or multistate device in which the load will change during a complete cycle. Examples of these types of loads would include the typical, agitator-style, gear-driven washing machine, and many modern-day television sets. The washing machine has multiple components, each with a contribution to the energy demands of the overall machine. Whenever filling the basin with water, only a level sensor is employed needing only small amounts of electricity. However, once filled the agitator motor begins, using a different amount of energy, and then during the spin cycle, a small pump motor drains the basin while the whole basin is spun, demanding even more electricity.

Another multiple-stage device in the home is the newer television set. When it is off, it really is not off. Five percent of the total electric current is still present in the television set, constantly charging capacitor components for precharge purposes. In this manner, the television set is "ready to go" and warmed up whenever the on button is pushed. The picture tube delivers an image quicker this way. Obviously, though, there are multiple load changes from a complete on state, to the off precharge state, to the full off state whenever the television is unplugged. There may be other electronic appliances within the home that function like this—having a trickle charge present even when the device is perceived to be off. There is a certain amount of debate as to whether this precharge convenience is worth the continual 5 percent waste when the device is not in use.

The most complicated type of load to monitor within the home is the continuously variable type. These types of loads are represented by dimmer switches, variable-speed tools, computers, high-efficiency lighting ballasts, and other rectification equipment such as the variable-frequency drives emerging in home products. The devices that produce these variable loads operate on the premise that the ac power being supplied to them is converted somehow into either dc or some reduced value of the actual power entering the device. What typically happens is that some energy is not used to operate the appliance or product, and the load appears to be varying. Often this unused energy is lost as heat. For example, the common dimmer switch often feels warm because full voltage is entering the dimmer but only a portion is allowed to go out to the light or fan. It may appear that energy is being saved, but actually the unused energy is being lost as heat.

2. *Automatic zone control.* The home, depending on the size, is usually divided into multiple zones. One zone typically might be the sleeping ar-

eas (bedrooms, bathrooms, etc.), and the other is usually the living areas (living room, family room, kitchen, etc.). These zones exist as a means to control the heating and cooling of certain sections of the house. Why waste energy heating or cooling certain areas of the house when those areas are not occupied? An unzoned house's heating or cooling system will deliver conditioned air everywhere that the ductwork runs even when there are no occupants in those areas. Thus twice as much energy is used for the home's heating and cooling. For this reason, zoned systems are practical. However, there must be an override to any zoned system. Occupants might be in the bedroom during the day and in the kitchen through the night. Therefore, an automatic zone control system is most practical and very attractive. Such a system typically works in this fashion: The home is still divided into multiple zones; however, each is automated by using motion detectors to alert the intelligent system that someone is occupying the room. This basically arms the heating and cooling system. If the zone thermostat calls for heating or cooling, then it is provided, and the occupant is comfortable. With this type of system, the room often heats or cools faster because the energy is directed to the occupied room versus the unoccupied room(s).

Multizone heating, ventilating, and air-conditioning controllers are now available to provide high-efficiency systems within the home. By individually controlling all the separate zones in the house, temperatures are economized. Each zone's thermostat and the heating/cooling system's dampers are under complete control of the multizone controller. Outdoor temperature also can be watched so as to optimize internal systems and determine a heat pump's lockout, or inefficient, point.

3. *Discrete on/off remote control.* Many electrical devices in the home can be left on unintentionally, and unless their on state is detected at some point, they can use a substantial amount of electricity over time. A prime example is the room lamp or light left on in a closet, or basement, or bathroom after an individual has entered and exited that room but neglected to switch the light to the off position. An intelligent home automation system could be programmed to monitor all the lights in the house along with all the outlets. One method is by a timer function built into the system master program (triggered in some way to "know" that the light is not on for a reason), which will automatically switch the light or appliance to the off position. Another method is to offer a full status map of any room or rooms in the house—which can be displayed on a main terminal(s). From here the viewer can, with a keyboard stroke, switch off the energy wasters from a central point.

4. *Direct control of appliances.* Just as many electrical circuits within the house often are dedicated to one appliance, as in the case of refrigerators, microwaves, and freezers, the intelligent home scheme can incorporate

that information into its program. This method of control might be nick-named "the forcing of on/off appliance states." As select outlets are defined throughout the house as nodes, complete with addresses, the main energy program can be set up to recognize where these appliances are within the house and actually control when they should come on or turn off. While refrigerators, freezers, and microwaves will determine themselves when to be on, many other appliances, when their location has been defined to the system host, can be programmed to come on at certain times of the day. This can be particularly cost-effective if electricity costs are lower late at night or during off-peak hours. Thus a loaded electric dryer can be programmed to dry the clothes at 2:30 A.M. if the electrical costs are much lower than during the daytime peak hours.

5. *Programmable control.* With virtually every appliance and light within the house addressable, the home owner can select which ones to turn on or off and when. Since every outlet has been wired for programmable control, this in turn allows the appliance plugged into that outlet to be controlled. This means that for a given condition, certain lights can come on, coffee makers can start brewing, or furnaces can be shut off. The user selects which devices he or she wants on by, in most systems, choosing from a menu of options. This selection process eventually becomes the user program for controlling electrical and energy events within the home. For example, you and your family are leaving in the early morning for the amusement park and know that you will be home late. It would be wasteful to turn on lights in the house and outside the house at 8:00 A.M. only to have them burning for several hours just so that when you arrive home at 10:00 P.M. you can see where you are going. A quick run through the menus can have inside and outside lights come on at, say, 8:00 P.M. Three hundred to four hundred watts of electrical power for 2 hours versus 14 hours is a substantial savings.

6. *Control by action.* This might be better defined as simply automatic control. If an action or event should trigger another action or event to happen, then this methodology should be used. This method more often could be employed when there is an emergency situation, but it also enters into energy-management schemes as well. Backup or redundant heating systems might be employed for complete outages of the primary system or simply when the weather is too severe and the heat pump is not an efficient choice. With severe weather, the heat pump could run constantly or its electric coils could remain on, soaking up valuable energy dollars. This is an instance where a backup gas-fired package could be useful. Likewise, when the weather is milder, an automatic decision to go with the mild-weather-efficient heat pump is in order.

7. *Control by mode.* There are many instances where we set up the lighting and heating of the house for a particular event or happening. This

can be also called *mode control*. We have such conditions whenever we end the day and go to sleep; we turn off the lights, maybe set the thermostat back for the whole-house heating system, and make sure that everything is secure. We may have a completely different set of conditions whenever we leave the house for a vacation. Some lights may have to be left burning all day long in order to make the appearance that someone is still at home while we are gone, or else a hard-to-program timer has to be located from the attic and set up for that light. In "vacation mode," the thermostat should be turned back or even the furnace shut off completely. This is a risky proposition because, depending on how long your trip is, the weather can change dramatically, for better or worse. If the outside temperature drops too much while you are away, the house plants may suffer or at worse maybe water pipes could freeze. This is why most heating systems should not be switched off during the winter months. You are enjoying that warmth on the beach, all the while your basement is flooding due to a frozen water pipe that has burst. An automatic home control system can be called on for a variety of modes—each with a prescribed set of rules for which it will guard, govern, and control the house's energy needs.

8. *Time-of-day control.* This method of control again uses the home automation system to maximize the efficiency of lights and appliances. Perhaps you will have to work late at the office and want the system to bring the furnace on just before you get home so that the house will be nice and toasty when you arrive. The system can be programmed before you leave for work, or even a call from work can adjust that arrival time, so that the furnace can save energy during the day while no one is in the house by not being on but simulate the action of someone physically moving the thermostat position up and therefore waiting 30 minutes for the house to warm back up.

9. *Setback control.* This is a very common method of energy management that can be accomplished in several different ways. The traditional, nonautomatic method involves walking up to the thermostat and moving the indicator back, or setting it back. Automatic systems have been developed over the years that accomplish the same functionality. One approach is to install an attachment to the existing thermostat that "fakes out" the thermostat into thinking it is warmer than it really is. A small amount of current is routed through this attachment, heating a resistive element that actually warms the immediate area at the thermostat, triggering its components. In this way the thermostat does not get the chance to close its contacts, thus starting the furnace cycle.

Another way to accomplish setback control is with a thermostat that can be set digitally from a host controller. This approach is more automatic, does not consume any "trickle charges" of electricity, and can be handled from a remote location. The drawback, however, is that the initial costs of a dedicated digital thermostat and complete home automation system have to be incurred. The concept is simple: Dial in the preset temperature that the thermostat is to track to, and when that value is detected, the furnace will be turned on. In this manner, nighttime and out-of-house conditions can be programmed automatically, and setback control energy savings can be realized—without having to get out of bed at midnight because the sound of the furnace coming on and off has awakened you from a good night's sleep (not to mention the fact that you cannot get back to sleep knowing that the furnace's thermostat setting is too high and you are throwing energy dollars away while you are trying to sleep).

10. *Communications links.* The beauty of an electronic home automation system is its ability to connect many pieces of hardware together and to communicate with a multitude of other computer-based systems. Because the basic premise of the automated home is to incorporate several microprocessor-based products, this approach lends itself well to sending and receiving electronic data between devices. Not only can a host computer control a great many independent appliances, but it also can "uplink" to other computerized pieces of equipment outside the home. With telephone and modem lines ever increasing in numbers in the average home, the communicating abilities of the present and the future are going to make this technology all the better.

Energy management's glue is the communications link. Without it, there really is no home automation nor enhanced energy control. The wires (or wireless systems) have to be run in order for all the independent devices to work together. Sensing equipment, "smart" appliances, host computers and controllers, and many other elements of the complete energy system have to communicate with each other. In doing so, energy savings can be maximized and even fine-tuned. A true system is multiple components working together, playing off of each other's best capabilities, and exploiting the technology.

11. *Remote access.* This method is actually a feature of the home automation system. It ties into the communication scheme and energy management extremely well. The thought or notion that comes to us concerning how we could be saving extra energy right this minute usually happens when we are in the car, at the office, or whenever we are not at the house. Intelligent energy systems allow us to telephone the host controller and actually change the program from a remote location. Turning the heat up or down is just a phone call away!

12. *Metering and monitoring.* This facet of energy management is another by-product of computer technology. Information and data are al-

ways useful, and by virtue of these computerized systems, great amounts of real data can be gathered. More important, power usage can be metered and monitored. Electrical information, watts used by an appliance, total current and load of devices, and other data can be gathered and used to determine where additional savings might be found. Communication links to the electric utility also can yield status reports for a given period of the month or month to month comparisons online and provide access to information on actual electricity costs before they get out of hand.

13. *Trending and analysis.* By virtue of being able to electronically record data, additional analysis of this information can now take place. Many application software programs are now available that can assimilate situations and predict energy usage and therefore pinpoint possible locations for energy savings. Perhaps during certain times of the day, electrical usage is dramatically increased. This may be an indicator that whatever activity is demanding the power needs to become a candidate for "energy scrutiny." Maybe the event or events can take place at nonpeak times of the day or night, thus reducing real costs. Maybe the appliance or device is old and/or wearing such that it is drawing too much electricity as compared with its original nameplate. This trending can pinpoint a problem area, an "energy glutton," and lead the home owner toward replacing or upgrading the device. The monitoring of usage, metering, trending, analysis, and overall understanding of the entire home "energy map" are important to the correction of any problem areas that may exist. We have to know where to start when it comes to energy management—and computerized energy systems allow us to do just that.

Energy-Consuming Products

Once we have determined the methods required for our control, comfort, and maximization of energy savings, it is time to get down to details. What products and appliances need to be factored into the equation? Which devices are the biggest energy users, and which should get the utmost in attention? A map of the house, an energy-usage floor plan, will help us to plan out a strategy. Whether building new or remodeling, this energy planning is a necessary step in optimizing the electricity and other energy sources entering the residence. It is often too easy to generally categorize the major energy appliances in the house as "the culprits" and not pay attention to all the lesser devices. *All electrical products use power. It all adds up!* Therefore, it is important to look at the entire picture—all the devices—and place a value on them.

By planning the home automation system around the appliances and equipment desired by the home owner, the system is much easier to wire, program, and control. Knowing what energy is required and where makes installation, and eventually, operation, much simpler. How often has it been said, "It is much easier to make the change on paper than it is to have to make the change on the actual job site with expensive labor (being paid by the hour)." Advanced planning will always pay for itself even if only allowing the home owner and home builder to avoid one small change (which could be expensive and time-consuming). This advanced planning is also another reason to have the home owner and home builder sit down and discuss the project. Upfront communication between groups is always productive.

Home Automation Tie-ins

Besides being ideally suited to keep energy costs in check, a residential automation system allows for many other disciplines and products to be connected to each other. From an energy-control standpoint, the home automation system can be integrated to control all appliances, the washing machine and dryer, the furnace and cooling equipment, all the lights, and automatic shutters or blinds, and even some window and door operation. Besides the energy-related control, the home automation system can tie in many other household components, including the security, audio, and video systems. By connecting these systems into the overall home controller, even more flexibility in energy efficiency is gained. For example, a security system can let us know if a door or window is open whenever the air conditioner is running. How often have we had to scurry around the house to manually close all the windows when the air conditioner comes on?

Besides allowing access to many other appliances and pieces of electronic equipment, a home automation system allows the residence to tie into or go online with many entities external to the home. Telecommunications advances are now allowing these "intelligent" systems to connect to the electric company or natural gas provider to monitor usage or to download some specialized energy-savings software from a host system. Home owners with residential automation systems also can be kept updated with Internet and e-mail postings concerning them and their systems. The home automation system can perform the automatic dialing and retrieval of messages from many sources for later review. Who knows, maybe in the future these systems will get to the point where they make even more decisions for us (artificial intelligence?).

Gas Systems

Many homes use natural gas as the energy source for heating, cooling, cooking, and hot water. These homes have two main energy sources to monitor. Unlike totally electric homes, the natural gas supplemental energy for the dual-supply homes has to be managed. Fortunately, the home automation industry has recognized this fact and has seen many suppliers of gas-related products make their devices more compatible with an electronic partner. After all, gas is energy, and energy conservation and management do not discriminate.

Several new product innovations have been incorporated into gas systems in the past few years. These innovations not only make them more readily usable with home automation systems but also allow for easier retrofitting and installation throughout the house. Many modern gas appliances are now provided with connectors that make installation convenient. Additionally, simple, more standard mating devices now can be incorporated into the house gas piping system. Rather than having to go through extensive work to match piping and connectors, the newer gas appliances are virtually ready to plug in. Thus, as is the case with electric appliances, it is merely a matter of plugging in and gas is now ready to be delivered to that appliance.

Conventional gas systems in the past typically were bound by the installation of black iron pipe or lack thereof. Heavy-gauge black iron pipe has no flexibility whatsoever and had to be put in place during the early stages of a building's construction. This also made retrofits very difficult whenever conversions from electric to gas were considered. Today, new semirigid stainless steel piping can be used in place of black iron pipe. With this more flexible material, gas service can be delivered to many more locations within a house, thus making gas appliances more attractive.

Perhaps the most attractive feature to using more gas appliances in the home today is their ability to be incorporated into the home automation scheme. This results not so much from the fact that each gas-powered appliance has integral intelligence and is able to connect up to a host controller (although some gas appliances are being supplied this way) but from the overall multiple-appliance control scheme. An intelligent gas subsystem, electronically connected to the home's main controller, can be installed with several gas outlets. Each gas outlet can be connected to a gas manifold that is fully controlled by the system. In this way, fewer fittings and pipe joints have to be installed for the given number of gas appliances used. What this means is that the chance for dangerous gas leaks is greatly minimized, additional gas appliances can be added without much difficulty, and a complete gas energy management system can be achieved and tied into the home's overall energy system.

Basic Hardware Components

The automated home achieves its status by virtue of all the interconnected components working together as a system. The heart of the system is the main controller, and as with any computer system, hardware selection is paramount in the ultimate performance of the package. From the main controller, its capabilities, power, and future expandability to the appliance themselves, all hardware components have to be selected carefully. Home automation consultants are emerging as professionals who can assist the home builder and home owner in making appropriate selections, but with enough research done, the average home owner can accomplish the same.

The first matter to address is the stage or state of the home. Is it in the planning stages, or have you just had an afterthought? Are you remodeling, or are you considering future needs as your children grow older? The physical state of the house's walls will dictate quite a bit as far as the direction and selection of hardware are concerned. As this industry evolves, the ability to work with existing, finished rooms is one criterion that home automation manufacturers are addressing. Location of appliances due to ease of wiring and installation should not always override convenience and practicality. Perhaps the most important aspect of the home automation system is the wiring. The wiring types, the methods of installation, and the peripheral issues surrounding the use of wiring in the home are extremely important—even more so than the host controller's functionality.

In commercial and industrial buildings, many different types of wires and cables are used for building automation and energy-management systems. This is no different with respect to residential automation and residential energy systems. There are many types of wire, and the future needs of a home's wiring are difficult at this time to predict (it is said that you can never overwire a house). However, locating the wire where it is not practical and installing the wrong type or gauge of wire present just as many problems as not having enough wiring.

For any automated energy system to be successful, proper power cable and lower-voltage control wire must be installed. A typical house's wiring, at a minimum, consists of the 110-V ac and 220-V ac electrical power wiring, telephone wiring, cable television (coaxial wire), and some low-level control wire. Using this list, we can build on our system's requirements. First, the electrical wiring that ultimately powers most of the appliances in the home must be installed properly by certified electricians. This wiring is very critical to the energy savings of the home. Connections at terminals and junction boxes should be solid so as to avoid heat losses resulting from loose connections. The gauge of the wire should be according to local building codes for the circuits that the wire is being dedicated to. All electrical wiring practices, circuit breakers, switches, and so on should be in accordance with the *National Electrical Code* and local building codes. Shortcuts and improper cost cutting can result in further energy losses and even fire. Follow the rules and regulations for your area, and select a qualified electrician.

Over the past 30 years, other wiring has become standard within the home. The telecommunications industry has helped pave the way for much of this new low-voltage wiring. Homes were wired initially for a single telephone. With this need came underground trunks of multiple low-voltage wires in huge bundles. As more and more telephones were installed, the telecommunications industry learned many things about capacitance, impedance, and electrical noise. New methods and types of wire had to be developed so that a person's voice, changed to an electrical signal, could be fully received on the other end once transduced back to a sound from that electrical signal. The telecommunications industry found that by twisting a loop of two wires they could eliminate unwanted electrical phenomena and maintain the integrity of the signal. This approach, called *twisted pair*, has become a standard in industry and most commercial facilities around the world. Today, the need for more than one telephone within the home has become the norm. Beyond telephone requirements, facsimile (fax) machines, modems, and the Internet have emerged as the next generation of "wired" components. If history is a good teacher, then the next 10 to 15 years are going to further change the home's base requirements for communications and control. Somewhere in this "spaghetti" lies the infrastructure for a complete energy-management system linked to devices within the home and outside.

With electrical power wiring basics set at one end of the residential wire-types spectrum and telephone wire systems at the other end, the remaining types of control wire fall somewhere in the middle. There's audio (speaker) wire, security and alarm wire, and computer wiring systems. The "intelligent" house uses all these types and then some depending on the manufacturer of the home automation product. Additionally, the *National Electric Code* (NEC) must be consulted whenever control wiring in the home is concerned. The NEC has placed some guidelines on wire types and methods to minimize dangerous conditions such as wire casings melting, fires, electric shock, and other electrically related hazards. Various articles in the NEC describe voltage and current limits that wire and cable can withstand. In cases of a full short circuit, the wire's insulating material is classified by a voltage and wattage maximum rating. Checking with local codes and the NEC will be well advised whenever contemplating special service wiring within the home. Usually, class 2 wire is used throughout the United States.

There are also defined categories or levels of control wiring that are used as guidelines for expected electrical performance. The levels typically are specified from 2 to 8 pairs in a bundle and relate to voice or data-application requirements. It is suggested that bundles containing 10 or more pairs can lead to electrical wave phenomena and signal degradation. The levels break down in this manner from highest performance to lowest:

Level 5: High-speed local-area networks (LANs) rated for 100 Mb/s

Level 4: Extended-distance local-area networks rated for 20 Mb/s; also for any level 1, level 2, and level 3 installation

Level 3: Medium-speed local-area networks rated for 15 Mb/s; also for any level 1, level 2, and level 3 installation

Level 2: Low-speed local-area networks rated for 5 Mb/s; also for a level 1 installation

Level 1: Basic analog and digital transmissions such as telephone and other low-speed installations

As more and more data are expected to travel across these electrical conductors, the need for the data to be received without interference has spawned a new requirement. This requirement is that known as *shielding* and provides an additional layer of protection against electrical noise. This electrical noise can come from several sources: electromagnetic interference (commonly called EMI), capacitive buildup in tightly bundled power cable, bigger electric motors in the home, light dimmers, compressors, fluorescent light ballasts, and "hash" or crosstalk. As more and more household equipment is becoming extra sensitive (due to microprocessor content), the need to "clean up" the electrical signals into and out of the device is critical. Shielding can provide this extra protection. It does so by encapsulating the current-carrying conductor with a mesh jacket or foil that must be grounded at one end. This practice should take unwanted noise away from the transmitted signal and keep its integrity high.

Shielded wire is obviously going to be more expensive, but as more and more information is passed from electronic product to electronic product within the home, shielded wire could become the norm. However, measures can be taken when routing control and signal wiring that does not carry shielding. One important method is to not run low-voltage signal wire with higher-voltage power wire. Keep telephone and data wires away from higher-current-carrying conductors. In doing this, eddy current and inductive phenomena can be reduced. Also, if wires have to come near one another, cross them at 90-degree angles of one another. This will minimize the instances that conductors are parallel to each other, and thus unwanted electrical exchanges or buildups cannot occur.

Twisted-pair wiring and shielding techniques will become more prevalent in the future. Today, the need to run two to four twisted pairs in a home is almost mandatory. Two-way communicating thermostats and sensors need this special wiring now, and the tremendous growth in computer, fax, and other communication line wiring is increasing this need. As we move into the next millennium, simply having a telephone line wired into the home for voice communications will be a negative. Multiple lines and types of lines are going to be needed in the house of the future. Besides twisted-pair wires and shielded wiring, other types of gangs and bundles are emerging.

The home automation system is a programmable control system that has the ability to perform certain actions under certain circumstances. Obviously, a core of software is embedded into the system controller that runs the home automation system. However, information must be provided to this core package with which decisions can be made: what to turn on, when to turn it on, how will the system know which device to turn on, etc. As a residence is built or retrofit with intelligent controls, the control system initially must be set up, or configured, for use with that particular house. Typically, every major appliance, electrical outlet, and many of the lights have to be identified— identified to the controller and its program. Thus each component to be controlled will be given its own "address." This unique address is a way for the controller to know to which location to send or receive information on the network. These addressable locations are called *nodes*, and this protocol is typical for home automation system networks as well as for most industrial or commercial systems.

Home automation energy systems require basic components in order to make a fully functional system. Every installation will be somewhat different, and a home automation specialist should be involved in the design and implementation of such a system. This involvement also should occur in conjunction with the builder and electrical contractor to ensure full compatibility. However, several components are common to most systems: the main controller, purchased either as a dedicated system controller or as a personal computer (retrofit to function as the home automation central controller), the conductor medium (or conductorless medium), receivers and transmitters, human interface panels, sensing equipment, and the appliances (ready to connect to a home automation system).

The "intelligent" house can be set up basically three ways: (1) running dedicated wire bundles to handle the home automation network, (2) using a technology that employs the existing house wiring as a medium for transmitting data, or (3) installing multiple infrared or high frequency receivers throughout the house (unfortunately, transmitters must be in close proximity to the receiver to get good, repeatable results). Alternatively, the house can be left as is, and the occupants can be the controllers.

Human Interface Panels

With the home automation program executing instructions behind the scenes, the home owner has to have a means of access to control that program. In other words, an interface between the human and the controller is necessary. As can be seen in Fig. 10.3, a sample screen from a control panel is shown. On this screen will appear several menu items from which the home owner can select. Different modes, settings, and conditions can be dialed in so that the system's computer functions around those entered. It is not necessary for the home owner to learn to program the system from a laptop or personal computer. Also, many new commands and instructions do not have to be learned with a user-friendly, menu-driven operator panel. In this way the home owner can set up the home automation system from a cursory vantage point, and the system will take over from there.

■ **10.3** *Automation system's sample screen from operator panel.*

The human interface panel or terminal can be the home's personal computer or it can be a dedicated "box" placed on the kitchen or any other wall. There has to be a device or mechanism by which the human user can enter data, receive information, and program the system. This is just an intermediary between the user and the actual home automation controller, as can be seen in Fig. 10.4. Here, the user enters data, which in turn are interpreted by the controller (actually the home automation computer), and action is taken by the computer based on the data that were input by the user. As in any automated system, the human interface panel also allows for a full manual override in the event the system is not functioning properly.

Thermostats and Sensing Equipment

Many energy systems work on the basis of closed-loop control. Figure 10.5 describes this type of control in a single-line diagram. The most common

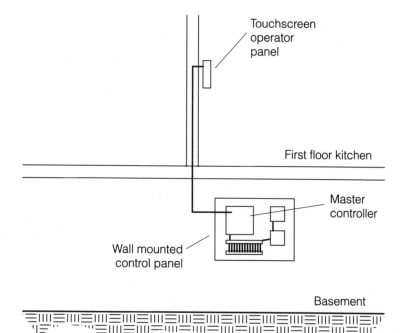

■ **10.4** *Operator panel attached to master or host controller.*

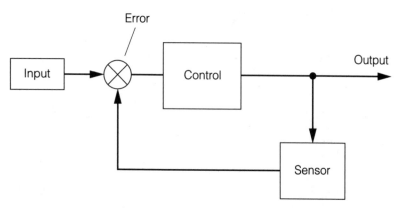

■ **10.5** *Closed-loop control.*

sensing element in the home is the thermostat, strategically located in the house so that it is triggered by changes in temperature. This trigger in turn actuates a furnace or air conditioner until the setpoint temperature is attained, thus triggering the thermostat again to shut off the heat-producing or cooling appliance. This is proportional, closed-loop control and is fairly straightforward. Automated energy systems use this concept and some more sophisticated thermostats to gain better control. In addition, zone and multizone controllers are being used to set different temperatures in different areas of the house.

A thermostat feature that gives the system a head start on especially hot or cold days is called *adaptive intelligent recovery*. It automatically calculates when to turn on the heating or cooling equipment to achieve a specific comfort temperature at the desired time. This maximizes your energy savings and comfort.

Besides setpoint temperature and zoned air climate control, automated energy systems are used to control the opening and closing of air/ductwork registers and dampers. Many different sensors are available for automated control today. There are other types of temperature sensors (e.g., a theristor, which is a variable resister, can sense values of temperature change), pressure sensors to sense static pressure changes in ductwork, air- and water-flow sensors that detect gas or fluid movement, and so on. Booster fans and whole-house fans can be incorporated into the control scheme and brought on whenever necessary. Variable-frequency drives are also finding their way into the residence as heat-pump, fan, and compressor controllers. More and more appliances will be found to be equipped with small microprocessors that will enable them to be networked to an energy system and thus yield better energy savings.

Electronic Wiring Systems

Electricity is the flow of negatively charged electrons from one point to another. These electrons travel by means of a conductor. A *conductor* is any material that allows for the flow or conduction of electrons. In a common wire, the metallic inner material is the *conductor*, whereas the outer casing, of a nonconductive material, is called the *insulator*. This is illustrated in Fig. 10.6. Most wire used today as the conductor is either aluminum or copper, soft metals that also can bend. The amount of aluminum or copper within a given plastic casing is called the *wire gauge* or *diameter*, as is shown in Fig. 10.7. This may be somewhat confusing because the wire gauge number is assigned to a certain thickness or cross section of the wire, and as the wire gauge number increases, the actual diameter decreases. Table 10.1 lists some common wire gauge sizes used in residential construction. For control purposes, the home owner and home builder will be working with mainly smaller-diameter wires (larger gauge numbers), since most of the time the voltages and current levels traveling through these conductors will be relatively low.

Today's and tomorrow's home owners probably will get more involved with the electrical wiring than they did in the past. This will be more evident as the quantity of appliances and electronic equipment increases within the home. Control wiring is also at a level where home owners are not afraid to

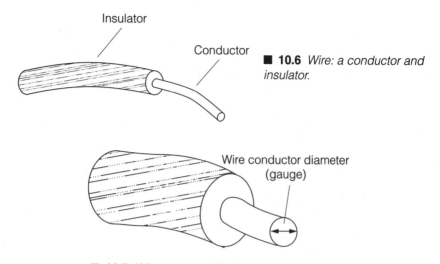

Insulator

Conductor

■ **10.6** *Wire: a conductor and insulator.*

Wire conductor diameter (gauge)

■ **10.7** *Wire cross section.*

■ **Table 10.1 Common Wire Gauges and Sizes**

Gauge	Diameter (in, not including plastic casing)	Comments
24	0.0201	Used as control wiring
22	0.0253	Used as control wiring
20	0.0320	Used as control wiring
18	0.0403	Used as control wiring
16	0.0508	Used as control wiring
14	0.0641	Residential 115 V ac wiring
12	0.0808	Residential 115 V ac wiring
10	0.1019	Residential 115 V ac wiring
8	0.1285	More than ⅛ in diameter

work with it (not to mention the fact that an expensive certified electrician cannot be called for every minor connection). When doing your own wiring, some helpful hints include the following:

1. When using wire, make it easy for the electrons to flow from point to point. Resistance is that property within a wire which opposes that flow of electrons. The electrons start colliding with each other and other atoms. This resistance shows up initially as heat in a warm wire. The ideal control wiring scheme should not have any warm wires.

2. Long wire runs or long lengths of wire have greater resistance and thus can get warmer than shorter runs of the same gauge wire. Always try to route wiring the shortest distance possible. If distances cannot be shortened, then consider a heavier-gauge wire. Shortening distances also minimizes the chance of electrical noise interfering with your control signals.

3. Thinner wires present limitations to electron travel because the electrons have less room to travel in the thinner wire. The potential for bumping into other electrons or other atoms is greater whenever the "highway" is small and crowded. Thicker wires give plenty of room for electrons to maneuver and provide "passing lanes."

As can be seen in Fig. 10.8, future wiring schemes will take on a much different physical look than those of 20 years ago. Actually, these wiring schemes are for today's residential automation systems. The more data that have to be transmitted, the greater the conductor's capacity has to be. For example, when downloading a text file, the time to download is tolerable, but when the file contains graphics, the size of the file becomes immense and the time quite lengthy. Thus a home's wire capacity will be a very important element in whether or not the home can handle new electronic technologies. Most probably, every electrical appliance within the home will someday carry a microprocessor in it, thus making it an electronic device. The home's wiring infrastructure had better be up to the task.

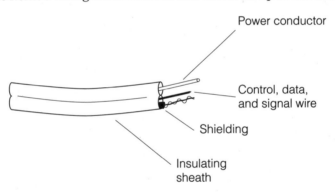

Power conductor

Control, data, and signal wire

Shielding

Insulating sheath

■ **10.8** *Future wiring schemes.*

Home Automation Wiring Types

Most homes have some wiring for basic electrical service. Some homes have more wiring than others. For instance, a home with 200-A service will have much more wiring than a home with 60-A service. This basic electrical power wiring is typically copper conductor wrapped in paper and coated with a plasticized insulation. There is also quite a bit of aluminum wire in many homes. Much of the wire is 12 and 14 gauge (the cross section or diameter of the wire). Appliances that require higher electric currents to operate usually have heavier-gauge wire to them. These include electric ranges, dryers, furnaces, heat pumps, and so on. Local codes determine the safe wire size for these appliances. The lower the gauge number, the greater is the cross section. A larger cross section allows more electrons to pass without extra "atomic collisions," and thus the electrical signal can be transmitted quicker. Figure 10.9 shows this phenomenon graphically. All

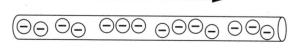

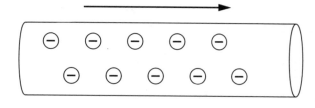

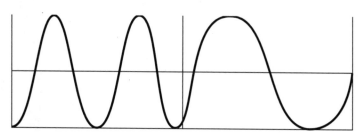

■ **10.9** *Electrons collide more often in thinner-gauge wire.*

electrical signals work on the same premise: Electrons flow through a conductor to a destination. These electrical signals can be higher voltages and currents to power appliances or lower voltages to send data. In either case, electrons are flowing. These are said to be *analog signals*. Sometimes data signals are *digital* in nature, which means that low-voltage pulses are sent through the conductor. Figures 10.10 and 10.11 show the difference between analog and digital signals. In the future, more digital signals will be sent over the home wire system than will analog signals.

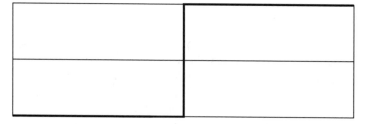

■ **10.10** *Analog signal.*

■ **10.11** *Digital signal.*

Some telecommunications companies are already anticipating tremendous growth in the need to move digital information, or data. The use of televisions, telefaxes (facsimile or fax machines), modems (modulator-demodulators for transferring data files), and basic telephone service is going to grow phenomenally. If a home's existing wiring is not able to handle the increased service, then transmissions will be slow and prone to interference.

Conventional electronic wiring systems might fall into the *category 5 type*, which means that several twisted pairs of high-capacity wire are bundled and used as the medium for most of the home's data and communication needs. However, fiberoptic cable is also a choice. Fiberoptic cable is glass-lined cable that allows laser-generated light pulses to flow through it. Voice, data, and graphic image signals can be transmitted virtually at the speed of light. This cable medium is capable of handling practically all of a home's future digital needs. However, this cable does have a higher initial cost when compared with other types. As is seen in Fig. 10.12, various bundles or wire groupings are compared relative to their composition and capacity.

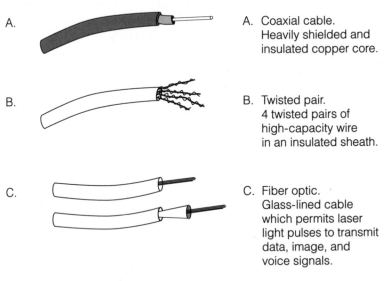

A. Coaxial cable.
Heavily shielded and
insulated copper core.

B. Twisted pair.
4 twisted pairs of
high-capacity wire
in an insulated sheath.

C. Fiber optic.
Glass-lined cable
which permits laser
light pulses to transmit
data, image, and
voice signals.

■ **10.12** *Wiring bundles.*

Wireless Systems

For decades, most homes have been wired the conventional way, some with more service than others. Unfortunately, the walls have been sealed and finished and are most likely framed such that adding special electronic wire is not going to be practical. Years ago no one could have predicted the electronic data explosion that was to happen and continues to proliferate. However, all is not lost for the homeowner with the desire to upgrade. There is the wireless system. While a hard-wired, dedicated electronic wiring scheme is much more effective (faster) and more impervious to electrical noise and interference, wireless systems are gaining in use, popularity, and capability.

Wireless systems fall into two groups: frequency-based and infrared. Many remote controls in the home today work on the principle of infrared operation. The television or VCR remote controls operate on this basis. Infrared systems function as rays of light, just beyond red in the light spectrum and which cannot be readily seen, are transmitted. These transmitted rays of light are received and decoded so that an occurrence can be initiated such as the turning on of an appliance. The fact of the matter is that virtually anything electrically actuated can be turned on or off by these remote controls. Unlike the hard-wired system, wireless control actually uses infrared signals or high-frequency radio waves, sent through the air, to start and stop the flow of electricity (ironically at a wired part) within another device. The frequency is the number of times every cycle a pulse in that radio wave strikes the receiver. There are 60 cycles every second typically. This means that there are an awful lot of pulses going through the air every second; 10,000 pulses per cycle, sometimes called 10 kilohertz (kHz), is actually low. High-frequency-based wireless remote control systems usually pulse at much higher rates than this (Table 10.2).

■ Table 10.2 Various Frequency Bands

Acronym	Range	Use
VLF (very low frequency)	2-30 kHz	Timing signals Industrial controllers
LF (low frequency)	30-300 kHz	Navigational
MF (medium frequency)	300-3000 kHz	Land, sea mobile
HF (high frequency)	3-30 MHz	Aircraft mobile
VHF (very high frequency)	30-300 MHz	Radio and TV broadcasting
UHF (ultra high frequency)	300-3000 MHz	Space and satellite communication
SHF (super high frequency)	3-30 GHz	Space and satellite communication

Notes:

1. The high-frequency (HF) range is also called *shortwaves*.
2. Audible frequencies are 30 Hz to 18 kHz (for younger people) and 100 Hz to 10 kHz (for older people).
3. The radio frequency wave that "carries" information is called the *carrier wave*.
4. kHz = kilohertz; MHz = megahertz; GHz = gigahertz.

Frequency-based wireless devices actually require two components, a transmitter and a receiver. The transmitter frequently is battery powered and converts that dc energy into high-frequency radio waves. These radio waves can be directed to a device that contains a radio-wave receiver that can decode the signal. A start, or on, signal from the transmitter will have its own particular frequency, whereas the stop signal will be at another frequency. These different frequencies allow for multiple functions in a wireless controller. The television wireless remote controller has many functions built into it. It can start and stop both the television and the VCR along with muting, channel, volume, and other control. Each of these functions requires a different wavelength or frequency be transmitted, then having it properly decoded at the receiving device, and then providing the result. If you have ever taken a remote control device apart (or had one fall apart), you will see that it is a static device (no moving parts). Its main components are the battery, board, and oscillator that develops the desired frequencies.

A technology that is a combination of wireless and hard wired is that called *X10*. It is a patented technology sometimes called the *power-line carrier*. It was developed in the 1970s in Scotland, and millions of these components are installed around the world. The technology works like this: Binary data are transmitted over the house's conventional power wires. This is a higher-frequency transmission made simultaneously with the 60-Hz ac sine wave. Figure 10.13 shows a typical ac sine wave over three cycles. The X10 technology uses a 120-kHz signal burst at the zero crossing point of the 60-Hz signal. Bursts of 1 ms are performed twice to reduce errors. In this way, a binary on, or 1, is a 120-kHz burst at the first crossing and none at the second, whereas a binary off, or 0, is no burst at the first zero crossing but a 120-kHz burst at the second. In this manner, a decoding device connected to the house's wiring system can know when to send and to receive messages. A message consists of multiple bits. Typically, the

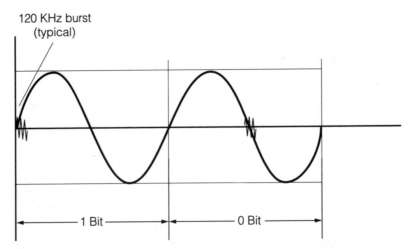

■ **10.13** *X10 transmits binary data with 120-kHz signal bursts at zero crossing points of household current.*

message is broken into a 4-bit start code, a 4-bit house code, a 4-bit function/unit code, and a 1-bit indicator bit. The indicator bit determines if the 4 previous bits are function or unit bits. In this way data can be transmitted all over the house in an orderly fashion. Every command is transmitted twice to ensure that the transmission is received "OK"; however, receiving devices only have to recognize the command once to operate. This is done to work reliably in situations where electrical noise has disturbed the system.

Sources of Electrical Noise

Just what the home owner needed—a new realm of potential problems. As homes get more acclimated to high-tech equipment and contain more wires and existing electrical systems and methods are employed, enter another by-product of the home automation technology—electrical noise. No, we are not about to discuss audible noise or loud-sounding equipment. Rather, unfortunately, every electrical circuit has the potential to have noise in it, and home automation systems are not immune to this noise. As a matter of fact, they are a noise-sensitive component that is more prone to this interference. A phenomenon that has been around for years with industrial and telecommunications markets, electrical noise is often hard to troubleshoot and diagnose. In these markets, computers and control systems operate on lower voltages, meaning that dips or surges in voltage can affect them. Electrical noise distorts the voltage and current waveforms of an electrical circuit in this manner. Home automation systems are comprised of the same basic elements—computers and controls.

As more homes are being equipped with computerized appliances and just more electrical service capability and wiring in general, the potential for problems due to noise is greater. *Potential* is a key word here, in that not every home system will experience this phenomenon, but it could. Trained electricians should take care whenever installing the wiring within a residence, but as more and more sensitive equipment gets installed, noise-related problems almost inevitably will surface. However, these electrical noise problems do have a solution, even though it may be expensive. A better understanding of what electrical noise is, its sources, and the things to be aware of during installations can minimize future problems.

A good example, which many have experienced, is the toy train set that runs usually only during the holiday season. Figure 10.14 shows a typical living room electrical circuit. Here, lights, television, VCR, and stereo equipment might all be running off the same circuit. Once the train set is plugged into that same circuit, certain devices exhibit signs of electrical interference, or noise. Frequently, the television picks up the electrical noise interference on the screen. What is happening is that the train's power sup-

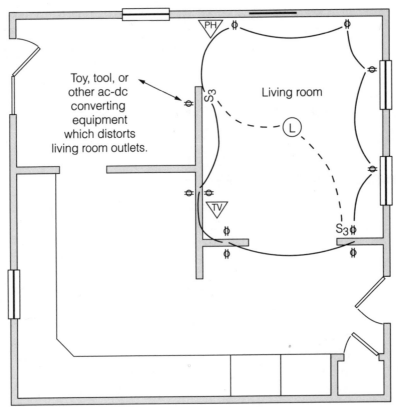

Toy, tool, or other ac-dc converting equipment which distorts living room outlets.

Living room

■ **10.14** *Living room circuit.*

ply is rectifying, or converting, alternating current (ac) into direct current (dc). This, coupled with the fact that the train's demand for usable dc changes as it goes from track to track, is what creates a disturbance on the circuit the power supply is on.

Another example that shows up on the television screen many times is the home computer. The personal computer is a device that is both rectifying ac to dc for internal use *and* is doing a lot of "switching" of this dc electricity internally (at incredible speeds—300 to 400 MHz these days). This can cause disturbance not only on the local circuit but also on peripheral circuits within the home. Thus the computer can distort as well as be affected by electrical disturbances, although the computer typically has some noise suppression built into its input power circuits. However, this is not always the case when a manufacturer builds a "cost-effective" microprocessor-based energy-saving appliance. These digital pieces of equipment often are more prone to voltage spikes (from lightening storms) and other noise disturbances on the circuit. The home owner may have to purchase extra surge suppressors and noise filters.

Fast and even large changes in voltage or current create electrical noise. This is the phenomenon being exhibited in the toy train and computer examples mentioned previously. There are other electrical and electromechanical devices that produce noise in a circuit. As more and more homes add this kind of equipment, the likelihood that these types of problems will surface increases. Wireless systems are not immune either. Disturbances from higher frequencies, radio-frequency interference (RFI), can be present in the home. More and more home equipment is falling into this category (remote telephones, walkie-talkies, computers, and even controllers) and either receives RFI or produces it. Thus noise sources include but obviously are not limited to:

☐ *Crosstalk* from power wiring to control wiring. Routing of high-voltage wire with low-voltage signal wire presents a condition for this. A good wiring practice is to cross wires at 90-degree angles to each other to minimize crosstalk.

☐ *Thunderstorms and lightening.*

☐ *Power switching or power conversion.* Fast-switching computer equipment and converting ac to dc can be contributors.

☐ *Fluorescent lights.* Many of today's fluorescent lights have electronic ballasts built in that can create minor disturbances. This is more noticeable the more of these types of lights are in service.

☐ *Ground loops.* As can be seen in Fig. 10.15, if equipment is not grounded properly, then a path for unwanted electrical noise is provided.

☐ *Electromagnetic interference (EMI).* Current through a conductor creates a magnetic field around the conductor. The strength of this field

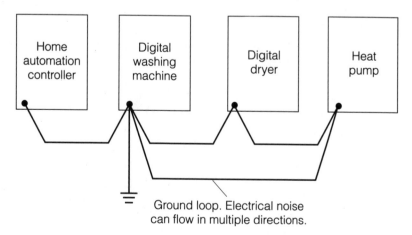

Ground loop. Electrical noise can flow in multiple directions.

■ **10.15** *Ground loops.*

is proportional to the amount of electric current flowing through the conductor. Sensitive wiring and equipment around this magnetic field can be the recipient of this unwanted noise.

☐ *Coupling of circuits.* This occurs where one circuit creates changes in another circuit through a common connection, capacitance (Fig. 10.16), or through induction (Fig. 10.17). This also can be through conducted (Fig. 10.18), radiated (Fig. 10.19), or shared impedance (Fig. 10.20) means.

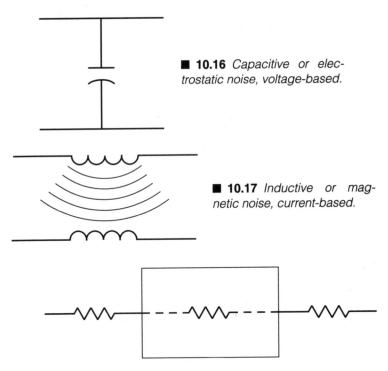

■ **10.16** *Capacitive or electrostatic noise, voltage-based.*

■ **10.17** *Inductive or magnetic noise, current-based.*

■ **10.18** *Conducted noise coupled into a circuit through wiring.*

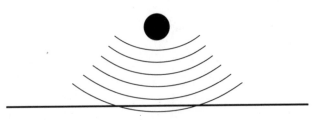

■ **10.19** *Radiated noise or radio frequency interference (RFI).*

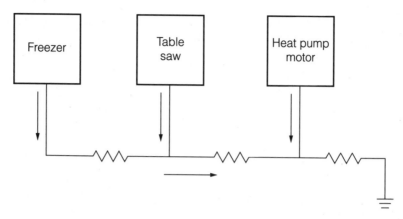

■ **10.20** *Shared impedance between pieces of electrical equipment.*

While most homes do not see these conditions routinely, the increases in electrical service and the equipment and appliance demands of the future continue to grow. We should recognize by now that we are very dependent on electricity and electrical equipment. We also should acknowledge that electrical service to every home is still on the rise (from 60-A service to 100-A to 200-A to 400-A service and beyond?). Future-proofing should include proper wiring practices so as to minimize noise-related problems later.

Application Software

The home automation system is really nothing more than some wire, "intelligent" appliances, some computer hardware, and sensing devices unless there is software to bring it all together. Just as good application software ("killer apps") had to emerge for the personal computer industry to "take off," so too does the home automation sector need good, sound packages. Not only are specialized packages necessary to coordinate control between appliances and sensors, but other programs exist for energy analysis and prediction. This software can be useful both to the home owner and to the home builder.

The home builder who offers an "intelligent" home automation system must select a package that is easy to explain, easy to use, and easy to support. The home builder's primary function is to build homes, not to debug software. Therefore, selection of the best package to offer by the builder is often based on what manufacturer offers the most components and can offer a full, complete system. A home automation supplier typically will offer the main control panel, the wiring bundles, the application software, and the supervision to make the installation a success. This ensures that compatibility should exist between components. If an individual embarks on selecting various individual system components (from different manufacturers), the interfacing to each component may not always be straightforward. The home builder has to be able to discuss the home control scheme with the prospective client and even be able to demonstrate how it works.

Likewise, the home owner needs a home automation system that runs in the background. The home owner makes decisions on how and why the system should operate and then turns the control over to the system. The home owner does not have the time or usually the ability to learn a complicated software package. Therefore, the home automation system must be easy to operate and extremely reliable. For example, if the home owner is going away for 3 days, all the system should need to know is the time period in which the house will be in the "away" mode. The home's temperature and other heating/cooling-related devices will be controlled accordingly. It should be this simple.

Another application of software is that of using it to estimate or predict energy savings. These packages are many and vary in capability depending on the area of the country one's structure is in, the cost of electricity, and any other energy-related components being used. However, with so much power built into computers and software programs today, packages that can provide both energy estimates and real-time, full control of a residence's appliances are emerging. A system can monitor electrical usage and sensing equipment and fully decide what to do—all in a matter of milliseconds. Thus energy savings by virtue of a home automation system constitute a dynamic system versus our old method of reviewing the monthly electric bill and stating that "we're using too much electricity, what are we going to do about it!"

Home automation software is a combination of many functions. It is an energy manager, discrete input and output controller, an event-control system, and a house "watchdog." Home automation software is comprised of multiple programs, invisible to the home owner, and all available to perform many household operations without intervention by humans. These software programs can monitor and control virtually anything in any manner, reacting to any condition or occurrence per your preset instructions. From an energy standpoint, here are some menu items that are available:

1. Remind you of upcoming dates/times regarding electricity or gas rates (peak and nonpeak). Also, if the system is connected via modem, can deliver energy-related messages for your review.

2. Allow you to call your home from anywhere in the world to turn a particular appliance on or off, check the status of any appliance, or increase/decrease the thermostat setting so that the house is comfortable when you get home.

3. Turn on your electric blanket on cold nights so that your bed will be warmed by bedtime.

4. Open the bedroom drapes, start the coffee maker, and open any shutters based on outside temperature.

5. Randomly turn lights on/off to make your home look occupied when you are away.

6. Monitor your energy consumption and predict your electric bill for the month.

7. Dim, turn off, or brighten lights to simulate different moods (or whenever a room is unoccupied for a period of time).

8. Control your heating and air-conditioning systems and water heater operation such that a comfortable temperature is maintained when you are home and energy is saved when you are away.

9. Save water by watering the lawn at night only when necessary.

10. Display status screens on the human interface panel, computer, or even the television screen.

11. Changes can be made via the human interface panel (or terminal), personal computer (if set up for this task), the telephone, radio frequency (RF) remote control, infrared (IR) remote control, and voice actuation. This allows the home owner to be able to carry out any desired task with the press of a button or by voice command.

The application software system should be a Windows- or DOS-based package with drivers and programs for analog and digital inputs and outputs, infrared systems, X10 technology, timers, voice recognition, telephone modem capability, and a user-friendly and menu-driven interface so that average home owners are not intimidated. As this technology evolves, many software packages will become available for both personal computers in the home and dedicated home automation controllers. The retrofit of existing personal computers for home automation use will be possible with the addition of the proper I/O (input-output) boards, software drivers, and other peripheral hardware.

Digital and Analog Input-Output (I/O)

Many appliances and electrical equipment and sensors have on and off states. This makes controlling them with a residential automation system possible. This is the most common type of I/O found in the automation environment. This I/O is either an "on" or an "off" state (sometimes high or low in terms of a voltage). A typical voltage rating for digital I/O can range from 2.3 to 5 V dc; thus a low-voltage power supply will be needed for any PC-based control system utilizing I/O. Sink currents for this type of I/O can be 24 mA per channel, whereas source currents will be 15 mA per channel. Digital I/O boards can be configured to have between 16 and 216 discrete inputs and/or outputs. Many computers use a passive backplane architecture with up to as many as 20 available expansion slots. Some remote I/O systems (racks where the physical wires are landed from the appliances) can be configured with as many as 2000 data points per serial I/O port. This number is quite high for the normal residence. Obviously, with multiple serial ports, quite an extensive system can be built from a single PC. However, one important tradeoff is that the more I/O, the longer will be the scan of the program and the longer will be the execution times.

Many of these plug-in I/O boards can have sampling speeds of up to 100,000 readings per second depending on the analog-to-digital (A/D) converter type and its resolution. By using a direct memory access (DMA) technique within the PC, throughput rates to system memory will be extremely high. For applications requiring a "snapshot" of transient events, an A/D board with sampling rates of 40 or 100 MHz should be employed. These cards can handle the short bursts of large amounts of data. It must be recognized, however, that serial interface will be limited to the standard baud rates available with these types of systems. Baud rates of 9600 to 19,200 are still the norm for serial communications. Thus a baud rate of the more common 9600 baud equates to approximately 960 readings, or points per second. As always, the central processing unit will have to handle other system functions besides the serial communications link, and thus the system will slow down even further.

Very similar to digital I/O, analog I/O has to be routed through some type of A/D converter before it can be analyzed by a home automation system. Analog signals have multiple ranges (0 to 10 V dc, 0 to 5 V dc, 4 to 20 mA, etc.) and thus can be scaled. Typical resolutions of analog signals can range from 8 up to 20 bits. Speeds that can be measured for analog I/O can be as high as 100 Mhz. Typical plug-in A/D boards will have 8 or 16 channels of analog input. Often, the analog board will come equipped with a few digital channels as well.

Analog and digital I/O boards have specific features. Many are multifunction cards, and you will pay for this functionality even if you do not need all of it. It sometimes is wise to take a moment and find the board that best suits your application. The most important is the I/O count, or how many and what type of I/O points you get with a one-card footprint. Many have 16 single ended, with 8 differential analog inputs. Twelve-bit resolution is common, as are 12-bit D/A or A/D converters (many boards will carry two) with switches on the board for the ranges. An analog output typically is disabled on power up and remains disabled until a user program writes to it. The other important feature should be that of how many samples per second (100K per second and more). Some boards come with 4 bits of transistor-transistor logic (TTL) or complementary metal oxide semiconductor

(CMOS)–compatible digital input capability, since these types of digital inputs are used commonly.

Another group of I/O tools consists of counters and timers. These useful components always seem to get classified together in the world of I/O. Used in a typical home automation or PC-based control scheme, they actually are intermediaries between the master controller and the I/O itself. The counter or, in ladder logic, counter blocks allow the programmer to count something. It may count inputs from a proximity switch so that it knows when it has met the predefined setting. Pulses from a pulse generator also can be programmed to be counted. Optical encoders can supply valuable pulse information about a variety of motion-control situations. By getting a trigger, or start, a reset contact, and being told how to increment, the counter block gets the home automation an answer more directly. The same can be said of the timer block. This block, while typically not needing a trigger from an external input but rather being engaged by the microprocessor itself, actually generates clock ticks to fully time an event. Timers and counters are often provided right on a multifunction analog and digital I/O card. The onboard internal clock is typically a 1.0-MHz clock, while access to an external clock (the main processor's) can be faster up to 10 MHz. Counters and timers often are concatenated to form a counter/timer of a higher bit count, such as 32, for timed A/D conversions and external frequency generation. These dual counter/timers are enabled by the home automation program and are clocked by a crystal oscillator source. Typical timer ranges should be no less than 2.5 MHz and not greater than 1 pulse per hour. Timing and counting functions within the home are going to be very useful for determining when shutters and drapes should open or close, when other electrical equipment should be on, and when to turn off the light in an unoccupied room. This is another great tool in the I/O toolbox available to the do-it-yourself home owner programmer.

There are several manufacturers of I/O products, home automation software, and controls. Many home automation manufacturers carry their own brand of I/O (and thus the end user needs the software driver for that I/O and usually can get it from the maker). The basic elements of the I/O hardware scheme are the actual modules, the rack to which the modules attach, and the other termination point. Once this arrangement is fully wired to all components, a functioning I/O system will be in place once the software recognizes what's out there. Different manufacturers of I/O have different color designations for input modules and output modules. Wire sizes are typically smaller gauge to accommodate the tighter terminals one has to wire to. Again, sound wiring practice in these situations is advised because only one critical I/O has to malfunction for a machine to go down. Luckily, most I/O packages come with light-emitting diode (LED) diagnostics to indicate a functioning I/O point.

The I/O used or the plug-in boards provided generally are supplied with the interface programs or samples to aid the programmer. These are often just called the *I/O driver* or *drivers*. In the case of remote I/O, the programming is done in ASCII command strings and should be fairly straightforward to use. Generally, remote I/O is programmed via the serial interface of the PC. Some menu-driven setup and calibration programs are provided as utility software for use with, for example, an analog board. Some drivers are supplied in a C language environment and as a linkable file.

PC-based home automation controls basically have to work with all types of I/O. Some home automation software packages come with drivers built into the source code, whereas other packages can accept or provide the I/O drivers needed. As to the selection of a particular package, thoroughly review the system with a person who will help implement it. Training, support, and the ability to upgrade and link to other intelligent devices are critical. This technology brings with it many questions, and the home owner, unless a computer programmer or electronics technician, will need answers—usually in a very short time frame. Telephone support and extensive training should be part of the package.

Home Automation Applications

As the home gets readied for the twenty-first century, energy management will become a secondary issue for many. With the automation system in place, we will transfer the energy-savings effort onto the "system." However, it is obvious that the future needs of our society will incorporate many other electronic services into the same energy-management system. Thus justifying the premium of installing this future-need hardware and software will be better realized when all other facets of electronic life get factored into the equation.

Besides the increase in telephone, fax, Internet, and modem services (how many homes presently have two or more telephones lines into them already), other services are needed. Television, security systems, audio/video systems, and other "intelligent" control systems are in need of a central control station within the home. The future also promises more bill paying, shopping, banking, and investment tracking by electronic means. The home's personal computer system will have to be upgraded to handle all this, the television system will become the control system, or a dedicated home automation controller will have to be incorporated into the residence.

The home of the future will be self-tuning. Yes, as it gets colder outside, the temperature on the inside will be adjusted to maintain a comfortable level. However, conventional temperature-control systems will not be in charge. A home central service center will be in charge. Multiple factors will be considered instead of the conventional single thermostat starting and stopping the furnace. Better control of the residence and its occupied rooms and more energy-efficient equipment will work in unison with house sensors, integral energy-saving application software, and home-owner needs to obtain a fully functional home energy system. To achieve the optimum in energy savings within any home, a home automation system has to be implemented. It will do the things you either forget about or do not have the time to handle.

The Future of Home Automation Systems

There most likely will be a couple of leaps in technology when it comes to automating the home. Just looking back over the past 100 years, we have gone from no electrical power to sending binary information over the ac line. Technological changes are occurring at astonishing rates in the fields of control, electricity, and microprocessors. Superconductivity and high-speed communication techniques are also emerging quickly. One technology typically can affect another, and thus major strides can be made.

As far as energy systems are concerned, the need to get the most out of our existing resources is mandatory. Energy systems and home automation actually help to conserve and minimize the waste of valuable energy. Until new energy sources emerge, we will continue using the nonrenewable resources that are available and convenient. This is why an elaborate, computerized energy system within the home is practical. As costs of components come down and more homes currently being built have "future-proofing" designed in, the use of these automated systems will increase. Once the issue of retrofitting the millions of existing structures and their wiring infrastructure is dealt with, this industry will proliferate.

Solar Energy

11

This chapter is about solar energy—how it works, and its uses. For hundreds of years, humans have been building homes and taking advantage of the sun in the design, orientation, and construction of those homes, in many instances not aware that they had even succeeded in doing this. Perhaps their builder did things in that fashion routinely, and the client was the beneficiary of a sound solar energy concept. The amount of sunlight that the earth receives each day alone is hundreds of thousands of times the total amount of electricity produced on the earth. Therefore, incorporating solar-related concepts into building designs now and in the future is a logical choice.

The idea of using the sun's rays to produce usable power on this planet is an old one. The process of capturing solar radiation and using it to produce steam on which machines could run has been around since the industrial revolution. However, since the world's energy needs grew much faster than the development of solar energy conversion techniques, solar projects never got much attention. All attention was focused on the search for fossil fuels that seemed to be easier to obtain and were able to keep up with the increased demand for energy. When Bell Laboratories developed the first practical solar cell in 1954, using solar energy for electrical needs got a boost. With development of the space industry, solar cells were used as a power source on its satellites. Also, as the various oil crises and oil embargoes happened in the 1970s, alternative energy sources came to the forefront. Being able to develop and produce silicon solar cells for industrial use became more attractive.

Solar Fundamentals

As the earth makes its elliptical orbit around the sun, its distance actually changes—closest in the winter and farthest in the summer—as can be seen in Fig. 11.1. The reason that this orbital phenomenon does not result in

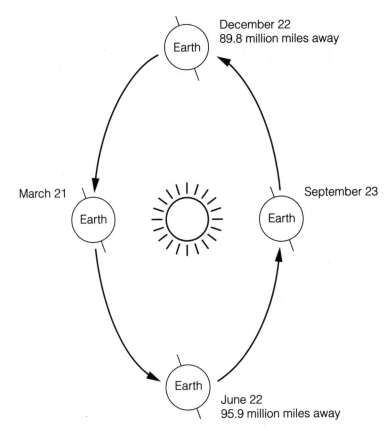

■ **11.1** *The earth's elliptical orbit around the sun.*

warmer winters and cooler summers is because of the earth's tilting on its axis. The north pole is tilted toward the sun in summer and tilted away from the sun in the winter. This explains the seasons of the year. Also, the sun appears to be higher in the sky during the summer and lower in the sky during the winter. It actually depends on the latitude at which the observer is measuring. However, all will agree that the infrared rays striking the earth every day are still worth pursuing and harnessing as viable home energy in some form or another.

Orientation is the most employed technique in getting the most from the day's sunshine. Typically, orienting the house toward the south yields the best heat gain. This means that windows or glazing surface areas should face this direction to pick up solar energy throughout the day. Additionally, thermal mass elements such as brick, concrete slabs, and block should be posed to soak up heat during sunny days. Even the color scheme should be coordinated for the southern exposure—dark, absorbing colors. It also should be noted that while focusing the house's orientation to the south, other important house components are not positioned haphazardly. Always know from which direction the prevailing winds and storms come. Garage doors should not face the windy side of the home, typically the west and north, because the winds will find their way through the doors and into the living quarters, thus diminishing heat.

Insolation

Solar insolation is the amount of solar radiation striking the earth's surface over a given period of time. It is measured in units called *langleys*, which are equivalent to one calorie of radiant energy per square centimeter. One langley is also equal to 3.69 British thermal units per square foot. Solar insolation should include other factors that affect the overall real value such as terrain, cloudy days, the air's moisture content, elevation with respect to sea level, and the overall lengths of days throughout the year. Trying to get data that are accurate regarding average daily solar radiation is sometimes difficult. The sources for these data provide a guide for the solar entreprenuer. More than likely if solar energy schemes are to be incorporated into a residence, then the home owner or home builder already knows that the region has the solar radiation basis to support such an endeavor. Maps do exist, however, that designate the prime solar radiation regions.

Because there are so many climactic conditions around the country and the world, solar energy for residential construction is very difficult to categorize and summarize. Yet there are basic principles that allow for an educated decision as to whether a solar energy system is appropriate for a given residential project. The purpose of this chapter is to provide a look at how solar energy can be used in typical residential construction without the need for high initial costs or full design around a solar scheme. In addition, this chapter looks at ways to incorporate into the home as many methods as possible to exploit the sun's energy.

Heat Gain

Earlier we calculated the heat loss for an entire residence in order to size the heating equipment. The *heat gain* of that residence was basically ignored with that calculation because design of heating systems has to be based on the worst conditions, those which would exemplify *no* heat gain on very cloudy, cold days or cold winter nights. However, every residence will have some type of heat gain, and this must be factored in when sizing the cooling system. Additionally, a home can be designed such that its overall heat gain is sufficient to actually heat the entire home or a portion of it for a high percentage of the heating season. The heat gain of a residence can be likened to an auxiliary heating system for the residence, and therefore, taking into account all the aspects of heat gain is worthwhile.

Heat gain not only is a factor of the amount of the sun's radiation collected and stored by the house but also is made up of other factors. Human occupants add a certain amount of heat to a building. While this is not a large amount of heat for a residence, approximately 350 Btuh per person, it still should be factored. This is why a large auditorium can get warm merely due to heat gain from the occupants. If 1000 people are in an auditorium, at 350 Btuh per person, there is an increase of heat of about 35,000 Btuh—this is substantial. Unless the residence will have numerous occupants, the heat gain from the human occupants is virtually negligible. There is also incidental heat gain from a building's lighting, electric motors, and refrigeration equipment. There is also some from cooking if exhaust fans are not used. However, the greatest amount of heat gain is from the sun's energy entering the house.

Glass types and shading contribute to the other sensible loads of the residence as far as heat gain is concerned. The glass content of the house may need to be considered most when orienting the house with respect to the sun. Glass can absorb at rates ranging from approximately 7 Btuh per square foot of glass for a northern-exposure window up to 70 Btuh per square foot of single-pane glass on a southern exposure. Obviously, these values are hard to calculate because the sun does not shine every day and does not shine in the same location on the house all day long either. Average values can assist in the estimation of heat gain due to windows.

Other elements of the house that contribute to heat gain are the walls and roof. The walls, if dark colored, will absorb more heat than will light-colored walls. Orientation of various dark colored walls will definitely affect the overall heat gain of the residence. While orientation is important for wall heat gain, it is not so important with respect to the house's roof. Roofs admit more heat than do walls because they face the sun constantly. Roofs are less likely to change color as are walls, which can be repainted often. Table 11.1 shows various building components and their representative contribution to the overall heat gain of a house. While heat gain is often not a major consideration in the design and construction of a house, it should be. As energy efficiency becomes a more important factor in the future of residential construction, so too will heat-gain calculations and the actual orientation of a building's windows and walls toward the sun.

■ **Table 11.1 Heat-Gain Multipliers for Various Building Components**

Building Component	Heat-Gain Multiplier (Taken Times the Square Footage to Get Btuh)
Glass, single, east or west, no blinds	85.0
Glass, double, east or west, blinds	44.0
Glass, double, north, blinds	14.0
Doors	11.4
Roof, concrete, 2 in of insulation	6.9
Walls, brick or stone	4.6
Walls, studs, insulated	2.9
Floor, concrete, 4 in plus of insulation	1.6

Methods Used to Employ Solar Energy for Heating

Passive solar energy systems typically mean that the residential structure must be designed with the sole intent of collecting solar energy. Traditional and conventional framing and building methods have to be altered to accommodate solar collecting schemes. These systems additionally have to collect this energy without much in the way of mechanical involvement. This would somewhat defeat the purpose of using solar radiation as an energy source only to expend other energy sources to transfer, collect, or otherwise use it. Therefore, passive solar radiation energy systems are a direct and nonmechanical method. Types of passive solar energy systems include thermal mass components, insulating covers and shutters, extra glazing in strategic locations, material and color optimization, and various shading techniques.

An active solar energy system uses external mechanical power to move the collected heat for heating or cooling purposes. This can be accomplished by the introduction of a circulating pump or fan to move any solar-radiated heat from one location to another. A good example is rooftop solar collectors; if they are not a trickle/gravity-type system, then a mechanical means must be provided to move the heated water to the house and rooms below. Integrated systems use technology to exploit materials and processes and tries to not use mechanical means to move the heat. After all, mechanical systems require energy, and that somewhat defeats the purpose. Backup systems may have to be in place for any of the aforementioned solar energy systems in case of severe weather and prolonged cloudy conditions. These backup systems may include oil burners, gas burners, wood stoves, and electrical heating equipment.

Methods Used to Employ Solar Energy to Make Electricity

While changing solar radiation into electricity is not a common occurrence with residential construction, the home owner and home builder need to be aware of this capability, especially for future needs. With deregulation of the electricity industry, home owners are now able, and will have even more flexibility in the future, to select from where they want to purchase their electrical power. This may mean that they can choose to purchase power from a solar energy source. Additionally, direct conversion of solar energy into electricity for home use is not that far off. New technologies and methods are being introduced that may make this function more practical for the home owner.

There are basically two ways that energy from the sun can be changed into electricity. One way is by solar thermal plants. In the solar thermal plant, the sun shines down on large mirrors that act as solar collectors. The mirrors heat a liquid running through them to a very high temperature, and this heated fluid is routed back to a boiler. Here, water is boiled to create steam, and this steam turns a turbine. These turbines in turn rotate generators that make the electrical power. Figure 11.2 illustrates how this entire process works to produce electricity from the sun.

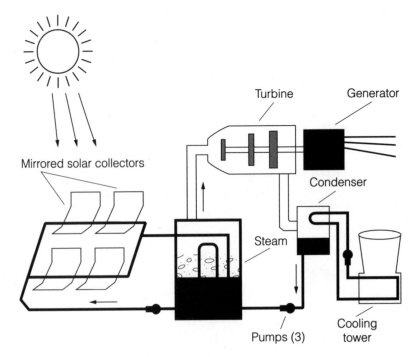

■ **11.2** *A solar thermal plant.*

The other way to create electricity from the sun is by using photovoltaic cells. These are special devices made up of a metal component and a semiconductor component. They are often referred to as *solar cells*. The inherent problem with this technology is that a great deal of them are required to make any substantial amounts of usable electricity. Also, on very cloudy days or, obviously, at night, the photovoltaic device cannot make electricity; therefore, backup systems, using fossil fuels again, have to be employed whenever necessary. However, there is growing interest in this technology, and many companies are investing in further research to develop the technology.

One advantage of photovoltaics is that there are no moving parts and the design is somewhat simple. There are no magnets, coils of wire, pumps, steam, or turbines—just many, many square feet of solar cells to get a small amount of electric current. We have all been exposed to the basic premise of photovoltaics in the solar watches and solar calculators that we see and use every day. Photovoltaic cells are composed of three base elements. There is a semiconductor material, in which there is a positive-negative junction, commonly called a *p-n junction*. Second, there is the metallic component, which is a grid structure that allows current to flow. The last component is a glass plate that allows the sun's rays to come in. This glass plate is accompanied by another plate that minimizes the amount of sunlight that will be reflected back by the cell. When sunlight hits the cell, the energy that is contained by the light induces electrons to jump from a lower energy level to a higher energy level within the *p-n* junction. Electric current can be defined as a movement of electrons due to a difference in potential, and this is what happens in these solar cells. The grid inside the cell functions as a transportation system for the electrons.

The cost of producing solar cells is presently not attractive to the consumer. With the added fact that so many are needed to get any appreciable electricity, photovoltaic technology is standing still. If the photovoltaic industry becomes capable of producing photovoltaic cells at a price of $.10/kWh of electricity produced, then the industry could explode. This is considering today's prices with today's other energy costs. If anything dramatic happens to the world's energy equilibrium (embargos, wars between vital nations, etc.), then the costs to produce photovoltaic cells will not be so unattractive anymore. One approach, called *spheral solar technology*, has been developed where more actual solar collecting surface area is provided on an equivalent per square foot basis as standard flat solar cells. This technology is shown in Fig. 11.3, but it has not really taken off as it was predicted to.

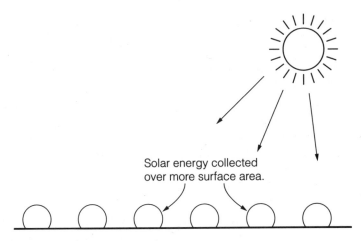

Solar energy collected
over more surface area.

■ **11.3** *The concept of spherical solar impregnated materials for more heat-absorbing area.*

Solar Energy Retrofits

Existing homes can be modified to use solar energy as a means of lowering heating bills. Just because the home is built around another heating scheme does not mean that remodeling cannot incorporate some established yet creative solar designs. First, it must be shown that the mean percentage of sun-filled days justifies, in your region, the pursuit of such an endeavor. Even though you may have lived in a particular region for some time, you still need to do a little research to check the history of sunshine in your region. Once this is done, check to see if there are any zoning restrictions in your region for solar collectors and solar-related construction. Next, find out if any state, local, or federal rebates (or funds) are available to help finance the project. Once you have done the advance homework, then selecting a worthwhile project can begin.

The addition of a sunroom or greenhouse is a very practical yet aesthetic way to gain energy from the sun. Sunlight entering the glass area can be trapped and used to help heat adjacent rooms of the house or even remote rooms (by mechanical means). Typically, sunrooms and greenhouses should be built onto the southern side of a dwelling. The excess heat can be ducted into other rooms, with the closest ones being the easiest to reach. The excess heat also should be available even after the sun goes down. This means that the construction, design, and materials chosen have to be thought out beforehand. Figure 11.4 shows an add-on greenhouse.

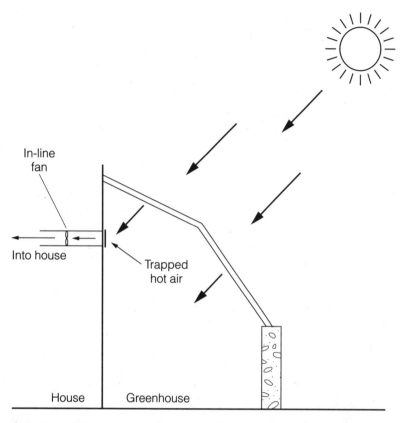

■ **11.4** *An add-on greenhouse and the introduction of that heated air into the residence.*

Not only can a greenhouse serve as a heat trap, exploiting the greenhouse effect, but it offers multiple benefits. It can grow plants and foods year round. By virtue of it being an addition to some exterior wall of the existing house, that exterior wall now is no longer an exterior wall. No more infiltration, concerns over insulation, and associated heat losses with any doors or windows that were in that wall. Additionally, since much of the remaining house is caulked and sealed to minimize (or virtually eliminate) air infiltration, the issue of the house being *too* airtight can be relieved by the presence of the greenhouse. It is relieved because the humidity and oxygen that plants give off are transferred with the ducted, warmed air. This means the house doesn't have to be stuffy, and the energy-conservation measures of making the whole house airtight do not come back to be a health hazard or cause other problems associated with homes being too airtight.

Solar Water Heating

While there are many individuals out there who have tried various techniques of heating by means of solar energy, there are four basic types of solar thermal systems. They are the direct system, the thermosiphon system, the drain-back system, and the indirect system. Each has its own features, and some are better used in harsher climates than are others. Direct and thermosiphon systems are generally used in tropical and subtropical climates. Drain-back and indirect systems can be used in cold climates where temperatures routinely go below freezing during the winter. This section provides a basic review; however, if the home owner or home builder desires to integrate a full solar heating system into residential design, then it is best to involve technical individuals who specialize in solar energy systems.

Direct systems work on the principle that the heat from the sun is transferred directly to the potable water in the collector. No antifreeze solution is used in this type of system. These systems are also known as *open-loop systems*. They are used most commonly in tropical and subtropical climates, where temperatures rarely go below 32°F (0°C). A solar collector, typically mounted on the roof of the structure where the water is to be used, heats the water. The hot water is stored in a storage tank. The storage tank, which is typically well insulated, is typically installed in the basement, garage, or utility room. Sensors are used to monitor the temperatures of the water in the system. A differential control unit attached to the sensors is used to control a circulating pump. If the temperature at the collector is 20°F greater than the temperature at the bottom of the storage tank the water in the system is circulated. When the differential of the water temperature is 5°F, the pump is turned off. In this way the water in the system is always being heated when the sun is shining. When required, a thermally operated valve, installed at the collector, is used to circulate the warm water into the collector as temperatures approach freezing. The system can be drained manually by closing the isolation valves and opening the drain valves instead of using a thermally operated valve if desired.

Thermosiphon systems are probably the least complex systems for use in solar heating. The use of thermosiphon, or free-flow, systems is accepted worldwide because of their simple and reliable characteristics. These systems have no pump or controls and are fully automatic in operation. Thermosiphon systems have their tank(s) positioned above the collector. As the water in the collector is heated by the sun, it rises into the tank mounted above the collector. This causes the cold water in the tank to flow into the collector, where it is heated. In this way, flow is created and the tank is filled with hot water.

Drain-back systems are used in cold climates to reliably ensure that the collectors and the piping never freeze. This protection is achieved when the system is operated in drain mode. In drain mode, all the liquid in the system's collectors and piping is drained back into an insulated reservoir tank when the system is not producing heat. An indicator on the reservoir tank shows when the collectors are completely drained. Each time the pump shuts off the solution in the collectors and piping is drained into the reservoir tank. The collectors are mounted at a slight angle to ensure that all the liquid can drain out of them. A solar collector, filled with either distilled water or an antifreeze solution, is used to collect the thermal energy in sunlight. Typically, distilled water or a propylene glycol and water mixture is used. The solution is circulated, by a pump, through the collector and into a storage tank, where a heat exchanger is contained. The heat exchanger transfers the heat into the potable water stored in the tank. The tank is designed to transfer the heat into the coldest water in the tank. The heat exchanger is a coil of tubing that wraps around inside the perimeter of the storage tank. The storage tank is typically installed in the basement, garage, or utility room and is well insulated.

As can be seen in Fig. 11.5, the very popular indirect solar heating systems are designed for use in climates where freezing weather occurs more fre-

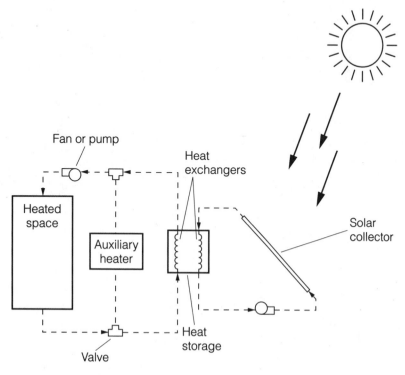

■ **11.5** *An indirect solar heating system.*

quently. These systems are also known as *closed-loop systems* and are used in cold climates where temperatures often go below freezing. A solar collector filled with an antifreeze solution is used to collect the thermal energy in the sunlight. Typically, a propylene glycol and water mixture is used. The solution is circulated, by a pump, through the collector and into a storage tank, where a heat exchanger is contained. The heat exchanger transfers the heat into the potable water stored in the tank. The tank is designed to transfer the heat into the coldest water in the tank. The heat exchanger is a coil of tubing that wraps around inside the perimeter of the storage tank. The storage tank must be well insulated and typically is installed in the basement, garage, or utility room. Sensors are used to monitor the temperature of the antifreeze solution in the system. A differential control unit attached to the sensors determines when the circulating pump should be activated. If the temperature at the collector is 20°F greater than the temperature at the bottom of the storage tank, the solution in the system is circulated in the closed loop. When the differential of the water temperature is 5°F, the pump is turned off. In this way, the water in the system is always being heated when the sun is shining. Indirect systems are closed-loop systems. There is no contact between the antifreeze solution and the potable water in the system.

When collecting energy from the sun and using it to heat a residence, it is usually viewed as counterproductive to use utility-supplied electricity to power the circulating pump. Therefore, it is sometimes desired or necessary to power the circulating pump and control unit with a photovoltaic system. The photovoltaic system, as discussed earlier in this chapter, converts sunlight into electricity. This electricity can be used to power any or all the electrical components in the system as long as there are enough solar cells. The photovoltaic panel or photovoltaic system and the components are matched in the design process to optimize overall system performance. In this way, a full, complete heating system is employed without the need for any energy except that provided by the sun's radiation.

Solar Facts and Closing Remarks

Solar energy has very little negative environmental impact. Solar energy does not create any smoke, soot, or pollution, as do coal- and oil-burning facilities. Solar energy does not require the mining of coal or the drilling for oil. The land is not ravaged, and animal homes are not disturbed. There is no radioactive waste to contend with, and it is a very safe energy source. The total solar flux that the earth receives daily is incredible. Yet most of it is not even used. The current contribution of solar energy to total energy production in the United States equals less than 0.5 percent. This is shown in Table 11.2. This percentage has got to increase in the future and probably will. The current manufacturing capacity of photovoltaic cells equals about 50 to 60 MW/year. This number, again, will increase in the future.

■ **Table 11.2 Percentage of Solar Energy Used in the United States (Units Shown Are in Quadrillion British Thermal Units, QBtus)**

Year	Total U.S. Energy Produced (QBtu)	Solar Energy Used (QBtu)
1990	70.7577	0.0669
1991	70.4203	0.0681
1992	69.9612	0.0677
1993	68.3164	0.0693
1994	70.6849	0.0685
1995	71.1194	0.0724
1996	72.6100	0.0749

Future Trends in Energy Systems and Controls

12

What energy strategies are going to affect the building construction industry in the future? Will new materials with super thermal insulating properties emerge? Will new methods overtake the old methods, or will the tried and true old methods survive? New techniques are being tried and applied every day for residential energy efficiency. Three key factors will change this industry: newly developed materials, microprocessor-based controls and appliances, and which energy sources will be most attractive from an availability and cost standpoint. The phrase "We don't do it that way anymore" may be used more often in the future. Electrical demands also will play a key role, especially as they increase.

Over the years, we have seen residential electrical demands continue to increase at a very steady pace. Sixty-amp service gave way to 100-A service, which gave way to 200-A service. It is not uncommon today to find services over 200 A being brought to some larger houses. As we add appliances and equipment, the need to increase wire size and actual electrical service has got to keep pace. This trend shows no signs of changing, as evidenced by the continual development of newer and better products, most of which are electrically based.

As for the future trends and developments, new materials that will offer tremendous insulating properties will emerge. Higher R values are appearing for a wide range of insulation products. Once a builder has success with a particular product, then the entire construction industry embraces it, and the standards committees check it out and eventually write it into their guidelines. Other new products also will be developed around existing products. For instance, steel studs have special formulations applied to them to help reduce heat transfer. Protrusions are fabricated onto the flanges of the studs so that a small air space exists between the stud and the finishing member. Similarly, holes can be inserted throughout the metal stud's length, further reducing heat transfer. This is not a new material but rather a new twist on an existing material—it is a new concept. The configurations of materials, different shapes, thicknesses, and sizes of homogeneous materials are constantly being experimented with and have an impact on residential construction. Every product being introduced seems to be digital in nature and equipped with a microprocessor. Washing machines, dryers, ranges, microwave ovens, televisions, VCRs, compact disc players, stereo equipment, and many more household appliances all have microprocessors these days. Not to mention the fact that nearly every household has at least one personal computer in it. All this computing power will give rise to a new realm of house control.

The home automation business is in its infancy. This market should explode in the early twenty-first century. New hardware and software are coming out every day for the specific purpose of controlling homes and making them as energy efficient as possible automatically. PC-based control has paved the way for the residential industry to boom. Industrial and commercial users of PC-based controls are providing a certain proving ground for this technology. Either using the home owner's own PC or installing a dedicated home controller, the future of residential life will be much more automated than it is today.

With the controllers and equipment being controlled all being digitally based, the problems of interfacing become less significant. Devices can communicate electronically with each other, and this, in turn, allows for full home automation control. We will not have to physically turn out the lights, lower the thermostat, or turn off the fan. The home automation system will sense when these occurrences must happen and see to it that they do occur. These intelligent house systems will lead us into even more fine-tuning of the energy-savings process in the residence. Considering the wasted electricity that the home automation system will detect and save and the extra savings that it will provide from monitors and sensors, this technology is a shoe-in for the future of residential energy systems.

New sensor technologies also should emerge to help save energy within the residence. Temperature, flow, and pressure sensors should become more affordable to residential consumers as the demand for these products increases. With pricing coming down, more automated systems will begin to use them. Without sensing equipment, the controller has to make decisions based on assumptions or on overall house temperature, etc. More temperature sensors in more areas of the house can provide better control. Additional sensor development to detect the presence of occupants within a room will allow for better and more efficient light control. New sensor technology and means of monitoring the electrical and energy systems of the house also will drive the home automation technology to more advanced levels.

Variable-frequency drives are just now emerging on home equipment. These devices continue to come down in price, and this means that they will become even more affordable to the home owner. When they begin appearing on more washing machines and more heat pumps, home owners, home builders, and manufacturers will find other residential applications for these energy-saving products. Every electrical ac motor within the home is a possible candidate for having a variable-frequency drive applied. Because they provide energy savings along with protection and better control of the electric motor, they are prime candidates to be fully integrated into any home automation system. Variable-frequency drives are microprocessor-based, which makes them attractive to these types of systems.

It is not possible to predict which energy sources will be most available and within the budget of the consumer in the next 10 to 20 years. We can speculate, but the world's economy and current events often dictate what happens to oil, coal, and gas prices. Projected energy costs have to have a big impact on the future of residential construction. In fact, builders have used this concern to provide the energy systems already in place in the buildings they erect today. Desperation and the need to keep warm on a cold winter's night often dictate what measures are taken to ensure building comfort.

Energy efficiency and energy savings require some sort of system to be achieved. Therefore, the science of energy systems is being formed. With so many components necessary to achieve the results of less heat loss, when to need more heat gain, and when to start and stop electrical equipment, it is no wonder energy savings as a field is so complex. If it were easy and simple, then books would not be written on the subject.

English-Metric Conversions

Area Conversion Constants

One square millimeter	is	0.00155 square inch
One square centimeter	is	0.155 square inch
One square meter	is	10.76387 square feet
One square meter	is	1.19599 square yards
One hectare	is	2.47104 acres
One square kilometer	is	247.104 acres
One square kilometer	is	0.3861 square miles
One square inch	is	645.163 square millimeters
One square inch	is	6.45163 square centimeters
One square foot	is	0.0929 square meter
One square yard	is	0.83613 square meter
One acre	is	0.40469 hectare
One acre	is	0.0040469 square kilometer
One square mile	is	2.5899 square kilometers

Weight Conversion Constants

One gram	is	0.03527 ounce (avoirdupois)
One gram	is	0.033818 fluid ounce (water)
One kilogram	is	35.27 ounces (avoirdupois)
One kilogram	is	2.20462 pounds (avoirdupois)
One metric ton (1000 kg)	is	1.10231 net tons (2000 pounds)
One ounce (avoirdupois)	is	28.35 grams
One fluid ounce (water)	is	29.57 grams
One ounce (avoirdupois)	is	0.02835 kilogram
One pound (avoirdupois)	is	0.45359 kilogram
One net ton (2000 pounds)	is	0.90719 tons (1000 kg)

Weight

10 milligrams	is	1 centigram
10 centigrams	is	1 decigram
10 decigrams	is	l gram
10 grams	is	1 decagram
10 decagrams	is	1 hectogram
10 hectograms	is	1 kilogram
1000 kilograms	is	1 (metric) ton

Length Conversion Constants

One millimeter	is	0.039370 inch
One centimeter	is	0.3937 inch
One decimeter	is	3.937 inches
One meter	is	39.370 inches
One meter	is	1.09361 yards
One meter	is	3.2808 feet
One kilometer	is	3,280.8 feet
One kilometer	is	0.62137 statute mile
One inch	is	25.4001 millimeters
One inch	is	2.54 centimeters
One inch	is	0.254 decimeter
One inch	is	0.0254 meter
One foot	is	0.30480 meter
One yard	is	0.91440 meter
One foot	is	0.0003048 kilometer
One statute mile	is	1.60935 kilometers

Length

10 millimeters	is	1 centimeter
10 centimeters	is	1 decimeter
10 decimeters	is	1 meter
1000 meters	is	1 kilometer

Volume Conversion Constants

One cubic centimeter	is	0.033818 fluid ounce
One cubic centimeter	is	0.061023 cubic inch
One liter	is	61.023 cubic inches
One liter	is	1.05668 quarts
One liter	is	0.26417 gallon
One liter	is	0.035317 cubic foot

One cubic meter	is	264.17 gallons
One cubic meter	is	35.317 cubic feet
One cubic meter	is	1.308 cubic yards
One fluid ounce	is	29.57 cubic centimeters
One cubic inch	is	16.387 cubic centimeters
One cubic inch	is	0.016387 liter
One quart	is	0.94636 liter
One gallon	is	3.78543 liters
One cubic foot	is	28.316 liters
One gallon	is	0.00378543 cubic meter
One cubic foot	is	0.028316 cubic meter
One cubic yard	is	0.7645 cubic meter

Energy Conversion Constants

One joule (J) = 0.738 ft•lb = 2.39×10^{-4} kcal = 6.24×10^{18} eV

One foot pound (ft•lb) = 1.36 J = 1.29×10^{-3} Btu + 3.25×10^{-4} kcal

One kilocalorie (kcal) = 4185 J = 3.97 Btu + 3077 ft•lb

One Btu = 0.252 kcal = 778 ft•lb

One electronvolt (eV) = 10^{-6} MeV = 10^{-9} GeV = 1.60×10^{-19} J = 1.18×10^{-19} ft•lb

Power Conversion Constants

One watt (W) = 1 J/s = 0.738 ft•lb/s

One kilowatt (kW) = 1000 W = 1.34 hp

One horsepower (hp) = 550 ft•lb/s = 746 W

One refrigeration ton = 12,000 Btu/h

Temperature Conversion Constants

$T = 5/9 \ (T_F - 32 \text{ degrees})$

$T = 9/5 \ (T_C + 32 \text{ degrees})$

$T = T_C + 273 \text{ degrees}$

$T = T_F + 460 \text{ degrees}$

Index

A

Absorption properties, 49, **49**
Absorptivity, 124
AC drives, 205–207, 225–244
AC motors, 211–214, **211**
Affinity laws, 216–217, **216**
Air changes, 162–163, **163**
Air conditioner, 142
Air locks, 130–132
Aluminum wire, 193, **195**
Amperage, 187–188
Amplitude, 190–192, **191, 192**
Apparent power, 185
Appliance system, 109
Application software, 275–277
Annual Fuel Utilization Ratio (AFUE), 141
Architectural specification sections, 20, **21**

B

Ballasts, 200–201
Biomass, 6
Black body, 124
Builder, 15, 16
Building wrap, 129–130

C

Capacitor, 196
Captive customers, 180
Choke, 198
Clean power, 183
Clients, 15, 19, 20
Coal, 8
Coefficient of Performance (COP), 154
Coefficient of heat transfer, 94, **98**
Cold, 3
Compressor cycle, 141
Condenser, 196
Concrete construction, 36, **37**
Condensation resistance factor, 57
Conductance, 94, **97**
Conduction, 4
Contractors, 15, 18, 19
Contestable customers, 180
Convection, 5
Cooling degree day, 108
Cooling load, 141

C (continued)

Cooling systems, 141–142
Coop, 180
Copper wire, 193, **194**
Crawl space, 91
Current, 187–188
Cycle, 191

D

Dead air space, 32, **60**
Dehumidifiers, 158–159
Demand, 177
Deregulation, 178
Diode, 197, **197**
Doors, 52–53, **53**
Drain back systems, 293
Dynamic Systems, 137

E

Economizer, 142
Effective power, 185
Efficiency, 138–141, **140,** 190
Electrical noise, 271–275
  capacitive, **274**
  conducted, **274**
  ground loops, **273**
  inductive, **274**
  radiated, **274**
  shared impedance, **275**
Electric meter, 175, **176**
Electricity, 183–184
Electricity broker, 181
Electricity cooperative, 180
Electromechanical, 138
Electronic homes, 247–248
Electronic passive solar systems (EPSS), 48
Emissivity, 48
Emissivity, 93, **95,** 124
Energy Policy Act (EPACT), 179
Energy saving tips, 62–66
Energy tracking, 119
Enthalpy, 139
Entropy, 139
Evaporation cooler, 141

F

Fans, 167–169, **168**
Film coefficient 93, **96**

Note: Illustrations are in **boldface**.

About the Author

Robert S. Carrow is the author of books on automation engineering for professionals and science project books for children. A widely respected expert in electronic controls who consults on solutions for industry, he wrote *Electronic Drives; SoftLogic: A Guide to Using a PC as a Programmable Logic Controller;* and *The Technician's Guide to Industrial Electronics*. He is also the author of two children's science books, *Put a Fan in Your Hat* and *Turn on the Lights from Bed*. All books are published by McGraw-Hill.